How to Repair and Maintain Sewing Machines

Troubleshooting, Fixing, and Servicing All Types of Sewing Machines for Beginners and Professionals

The Fix-It Guy

Copyright © 2024 by The Fix-It Guy

All rights reserved. No part of this book may be reproduced in any form or by any electronic or mechanical means, including information storage and retrieval systems, without permission in writing from the publisher, except by a reviewer who may quote brief passages in a review.

Table of Contents

Introduction

Chapter 1: Introduction to Sewing Machine Repair
- Why Learn Sewing Machine Repair?
- Safety Precautions and Essential Tools
- Understanding Sewing Machine Types and Brands

Chapter 2: Sewing Machine Basics
- Anatomy of a Sewing Machine
- How Sewing Machines Work
- Common Problems and Quick Fixes

Chapter 3: Vintage Sewing Machine Repair
- Identifying Vintage Models
- Cleaning and Oiling Antique Machines
- Restoring and Replacing Vintage Parts

Chapter 4: Modern Domestic Sewing Machine Maintenance
- Troubleshooting Electronic Sewing Machines
- Repairing Computerized Models
- Maintaining Plastic Components

Chapter 5: Industrial Sewing Machine Repair
- Understanding Industrial Machine Mechanics
- Timing and Adjusting Heavy-Duty Machines
- Solving Industrial-Specific Issues

Chapter 6: Serger and Overlocker Maintenance
- Mastering Serger Threading
- Blade Replacement and Adjustment
- Tension Troubleshooting for Perfect Stitches

Chapter 7: Advanced Repair Techniques
  - Motor Repair and Replacement
  - Circuit Board Diagnostics and Fixes
  - Tension Assembly Overhaul

Chapter 8: Preventive Maintenance and Care
  - Creating a Maintenance Schedule
  - Proper Cleaning and Lubrication Techniques
  - Storage and Transportation Tips

Chapter 9: Troubleshooting Guide
  - Diagnosing Common Issues
  - Step-by-Step Problem-Solving Flowcharts
  - When to Seek Professional Help

Chapter 10:   Resources and References
  - Recommended Tools and Supplies
  - Parts Sourcing Guide
  - Online Communities and Further Learning

Appendices
  - Sewing Machine Repair Glossary
  - Brand-Specific Maintenance Tips
  - Conversion Charts and Tables

Introduction

Have you ever gazed at your beloved sewing machine, frustrated by a stuck bobbin or skipped stitches, wishing you could diagnose and fix the problem yourself? Or perhaps you've come across a beautiful vintage machine at a flea market, imagining the potential it holds, but feeling daunted by the prospect of restoring it to working order? I understand those feelings completely.

Like many sewing enthusiasts, I used to be intimidated by the inner workings of sewing machines. Whenever my trusty machine encountered issues, I'd reluctantly consider expensive repairs or even replacement, even though I longed to understand and maintain it myself. But then I discovered the transformative power of mastering sewing machine repair and maintenance, and it changed everything.

Imagine being able to troubleshoot any sewing machine problem with confidence, from tension issues to timing troubles. Picture yourself skillfully servicing vintage machines, fine-tuning modern computerized models, and even tackling industrial powerhouses. With "How to Repair and Maintain Sewing Machines," you'll gain the knowledge and skills to do just that.

This comprehensive guide is designed to empower you with expert techniques, insider tips, and step-by-step instructions for diagnosing, repairing, and maintaining all types of sewing machines. Whether you're a hobbyist looking to care for your own machine or a professional aiming to expand your skillset, this book will take you from novice to sewing machine expert.

Here's what you'll discover within these pages:
- The essential tools and materials for successful sewing machine repair, and how to use them effectively
- Techniques for identifying common issues across various machine types and brands
- Secrets to achieving professional-level maintenance, from proper cleaning to precise oiling
- Step-by-step guidance for repairing and adjusting key components, from feed dogs to tension assemblies
- Expert advice on restoring vintage machines to their former glory
- Troubleshooting tips for modern electronic and computerized sewing machines
- Professional strategies for maintaining industrial machines and sergers

By investing in this guide, you're not just buying a book – you're unlocking the potential to keep your sewing projects running smoothly for years to come. You'll save hundreds, if not thousands, on professional repairs, and gain the satisfaction of understanding and caring for your machines with your own hands.

So why settle for unreliable machines or expensive repair services when you can become a sewing machine expert? Dive in now, and discover how to enhance your sewing experience, extend the life of your machines, and perhaps even turn your new skills into a profitable side business.

Are you ready to unleash your inner technician and start creating a more reliable, efficient sewing setup? Let's embark on this exciting journey together and unlock the secrets of professional-quality sewing machine repair and maintenance!

Chapter 1
Introduction to Sewing Machine Repair

Why Learn Sewing Machine Repair?

Picture yourself seated at your sewing machine, ready to bring your latest creative vision to life. You press the foot pedal, but instead of the familiar hum of stitches forming, you're met with a jarring clunk, followed by silence. Your heart sinks as you realize your trusted companion in countless projects has suddenly become an enigma of metal and plastic. In this moment, wouldn't you give anything to have the knowledge and skills to diagnose and fix the problem yourself? If you've ever found yourself in this frustrating situation, longing for the ability to breathe life back into your machine, you're not alone. And more importantly, you're in the right place.

As a sewing enthusiast or professional, understanding the intricacies of sewing machine repair isn't just a useful skill—it's a game-changer. It's the difference between a minor hiccup in your creative process and a complete halt to your projects. It's the power to transform that vintage find from a mere decoration to a functional piece of history. But why exactly should you invest your time and energy into learning sewing machine repair? Let's explore the compelling reasons that make this skill set invaluable for anyone who works with fabric and thread.

That's where our exploration of the benefits of learning sewing machine repair comes in. In this section, we'll uncover the many advantages that mastering this skill can bring to your sewing journey, empowering you with the knowledge to approach your machines with confidence and enthusiasm.

Essential Benefits of Learning Sewing Machine Repair

Mastering sewing machine repair offers a range of benefits that contribute to a more fulfilling and efficient sewing experience. Let's delve into these advantages in detail:

1. Cost Savings

Learning to repair your own machines can significantly reduce your sewing-related expenses:
- Avoid costly professional repairs for minor issues
- Extend the life of your current machines, delaying the need for replacements
- Save money by purchasing and restoring used or vintage machines

2. Increased Independence

Become self-reliant in maintaining your sewing equipment:
- Diagnose and fix issues without waiting for professional help
- Adjust and fine-tune your machine for optimal performance
- Customize your machine's settings for specific projects or fabrics

3. Enhanced Understanding of Machine Mechanics

Gain a deeper appreciation for the tools of your craft:
- Understand how different parts work together to create stitches
- Make more informed decisions when purchasing new machines
- Troubleshoot issues more effectively, even on unfamiliar models

4. Improved Sewing Quality

A well-maintained machine leads to better sewing results:
- Ensure consistent stitch quality by keeping your machine in top condition
- Adjust tension and timing for perfect stitches on various fabrics
- Reduce frustration caused by machine malfunctions during projects

5. Vintage Machine Restoration

Open up a world of possibilities with older machines:
- Breathe new life into high-quality vintage machines
- Experience the joy of sewing on beautifully crafted classic models
- Preserve sewing history by restoring and using vintage machines

6. Potential Income Opportunities

Transform your skill into a side business or career enhancement:
- Offer repair services to other sewists in your community
- Restore and sell vintage machines for profit
- Enhance your value as an employee in sewing-related industries

7. Environmental Responsibility

Contribute to sustainability through machine maintenance and repair:
- Reduce electronic waste by repairing rather than replacing machines

- Give new life to old machines that might otherwise end up in landfills
- Promote a culture of repair and reuse in the sewing community

8. Problem-Solving Skills

Develop valuable critical thinking abilities:
- Enhance your analytical skills through diagnosing machine issues
- Learn to approach problems methodically and creatively
- Apply these skills to other areas of your life and work

9. Community Building

Connect with other sewing enthusiasts on a deeper level:
- Share your knowledge and help others with their machine issues
- Participate in sewing machine repair forums and groups
- Organize workshops or classes to teach others these valuable skills

Considering Learning Sewing Machine Repair

While the benefits of learning sewing machine repair are numerous, it's important to consider a few factors before diving in:

1. Time commitment: Assess whether you can dedicate time to learning and practicing repair skills.
2. Tools and resources: Consider the initial investment in repair tools and learning materials.
3. Learning style: Determine whether you prefer hands-on learning, online courses, or a combination of both.

4. Machine types: Decide which types of machines you want to focus on (vintage, modern, industrial, or a combination).

Remember, while learning sewing machine repair requires effort and patience, the long-term benefits often far outweigh the initial investment of time and resources.

Understanding why learning sewing machine repair is valuable sets the foundation for your journey into this skill set. It affects everything from your daily sewing experience to potential career opportunities and your impact on the sewing community. By familiarizing yourself with these advantages, you're taking the first step towards becoming a more self-reliant, knowledgeable, and capable sewist.

In the following sections, we'll delve deeper into the anatomy of sewing machines, essential tools and materials, common issues and their solutions, and step-by-step instructions for various repair and maintenance tasks. Are you ready to empower yourself with the ability to care for and repair your beloved sewing machines? Let's continue this exciting journey together!

Safety Precautions and Essential Tools

Before we delve into the world of sewing machine repair, it's crucial to establish a foundation of safety. Remember, you're working with electrical components, sharp needles, and moving parts. Let's break down the key safety precautions:

1. Electrical Safety
   - Always unplug your machine before opening it up or performing any repairs.
   - Inspect power cords regularly for fraying or damage.
   - Use a surge protector to safeguard against power fluctuations.

Troubleshooting Tip: If you suspect an electrical issue, use a multimeter to test for continuity in wires and components before replacing parts.

2. Eye Protection
   - Wear safety glasses or goggles to protect against flying debris or broken needles.
   - Ensure good lighting in your work area to reduce eye strain.

3. Hand Protection
   - Use needle-nose pliers or tweezers when working with small parts to avoid pricking your fingers.
   - Consider wearing cut-resistant gloves when handling sharp edges or disassembling machines.

4. Workspace Organization
   - Keep your repair area clean and well-organized to prevent losing small parts.
   - Use a magnetic parts tray to keep screws and metal components from rolling away.

Troubleshooting Tip: Take photos or videos of your disassembly process to aid in reassembly later.

5. Chemical Safety
   - When using solvents or lubricants, work in a well-ventilated area.
   - Always read and follow manufacturer instructions for any chemicals used.
   - Store cleaning solutions and lubricants properly, out of reach of children and pets.

6. Machine Stability
   - Secure your machine on a stable work surface to prevent accidents during repair.
   - Use clamps or a non-slip mat if necessary to keep the machine steady.

Essential Tools for Sewing Machine Repair

Now that we've covered safety, let's explore the tools you'll need to become a sewing machine repair expert. Having the right tools not only makes your job easier but also helps prevent damage to your machine.

1. Screwdrivers
   - A set of precision screwdrivers in various sizes (both flathead and Phillips head)
   - A magnetic screwdriver for hard-to-reach areas

Troubleshooting Tip: If a screw is stuck, try applying a drop of penetrating oil and let it sit for a few minutes before attempting removal.

2. Pliers and Tweezers
   - Needle-nose pliers for gripping small parts
   - Bent-nose tweezers for reaching into tight spaces
   - Wire cutters for trimming thread or wires

3. Cleaning Tools
   - Lint brush or small paintbrush for removing dust and lint
   - Compressed air canister for blowing out hard-to-reach areas
   - Soft cloths for wiping down surfaces

Troubleshooting Tip: When using compressed air, always hold the can upright to avoid spraying liquid propellant into your machine.

4. Lubricants and Solvents
   - Sewing machine oil (never use WD-40 or other household lubricants)
   - Isopropyl alcohol for cleaning
   - Solvent for removing old, sticky lubricant

5. Diagnostic Tools
   - Multimeter for testing electrical components
   - Magnifying glass or loupe for inspecting small parts
   - Flashlight or headlamp for improved visibility

6. Specialty Tools
   - Tension gauge for measuring and adjusting thread tension
   - Needle insertion tool to aid in proper needle placement
   - Timing gauge for adjusting hook timing on more advanced repairs

Troubleshooting Tip: If you're unsure about the correct tension, test on scrap fabric and adjust gradually, keeping notes of your changes.

7. Organization Aids
   - Magnetic parts tray or small containers for organizing screws and small components
   - Label maker or masking tape for labeling parts during disassembly
   - Digital camera or smartphone for documenting the repair process

8. Reference Materials
   - Your sewing machine's manual (if available)
   - Repair guides specific to your machine model
   - A notebook for keeping repair logs and notes

Troubleshooting Tip: Create a "map" of screw locations on paper, placing each screw on the map as you remove it to ensure proper reassembly.

Building Your Toolkit

As you begin your journey into sewing machine repair, you don't need to purchase every tool at once. Start with the basics:
   - A good set of screwdrivers
   - Needle-nose pliers
   - Cleaning brushes
   - Sewing machine oil
   - A magnifying glass

As you gain experience and tackle more complex repairs, you can gradually expand your toolkit. Remember, investing in quality tools will pay off in the long run, making your repairs more efficient and effective.

Maintenance Routine

Establish a regular maintenance routine for your sewing machine:
1. Clean lint and dust after every few projects
2. Oil moving parts monthly or after 8-10 hours of use
3. Check and replace needles regularly
4. Inspect the power cord and foot pedal quarterly

Troubleshooting Tip: Keep a maintenance log to track when you've performed various tasks and note any recurring issues.

By following these safety precautions and equipping yourself with the right tools, you're setting a strong foundation for successful sewing machine repairs. Remember, patience and persistence are key. With practice, you'll gain confidence in your ability to diagnose and fix issues, extending the life of your beloved sewing machines and enhancing your sewing experience.

In the next section, we'll explore the anatomy of different types of sewing machines, providing you with the knowledge to tackle repairs on various models with confidence. Are you ready to dive deeper into the inner workings of these fascinating machines?

Understanding Sewing Machine Types and Brands

The world of sewing machines is vast and varied, with each type and brand offering unique features and challenges when it comes to repair and maintenance. Understanding these differences is crucial for anyone looking to master sewing machine repair. Let's explore the main categories and popular brands, along with their specific characteristics and common issues.

1. Mechanical Sewing Machines

These are the workhorses of the sewing world, known for their durability and straightforward design.

Characteristics:
- Fully manual operation with dials and levers
- No computerized components
- Often heavier due to metal construction

Popular Brands: Singer, Janome, Brother

Common Issues:
- Tension problems
- Timing misalignments
- Worn gears or belts

Troubleshooting Tip: When dealing with tension issues on mechanical machines, always check for lint buildup in the tension assembly before making adjustments.

2. Electronic Sewing Machines

These machines incorporate electronic components for enhanced functionality while retaining some manual controls.

Characteristics:
- Push-button stitch selection
- LED displays
- Automated features like needle threading

Popular Brands: Bernina, Pfaff, Husqvarna Viking

Common Issues:
- Circuit board failures
- Faulty LED displays
- Sensor malfunctions

Troubleshooting Tip: For electronic machines with unresponsive buttons, try resetting the machine by unplugging it for 5 minutes before troubleshooting further.

3. Computerized Sewing Machines

The most advanced domestic machines, offering a wide range of automated features and programmable functions.

Characteristics:
- Touchscreen interfaces
- Hundreds of built-in stitches and designs
- USB connectivity for pattern uploads

Popular Brands: Juki, Bernina, Brother

Common Issues:
- Software glitches
- Touchscreen calibration problems
- Memory errors

Troubleshooting Tip: Always check for and install the latest firmware updates for computerized machines to prevent software-related issues.

4. Overlockers/Sergers

Specialized machines designed for finishing edges and creating professional-looking seams.

Characteristics:
- Multiple thread paths (typically 3-5 threads)
- Differential feed for handling various fabrics
- Built-in knife for trimming fabric edges

Popular Brands: Juki, Babylock, Janome

Common Issues:
- Thread tension imbalances
- Knife dulling or misalignment
- Looper timing problems

Troubleshooting Tip: When adjusting serger tensions, always start with the lower looper, then upper looper, followed by needle threads.

5. Industrial Sewing Machines

Heavy-duty machines designed for high-volume, commercial use.

Characteristics:
- Powerful motors for continuous operation
- Specialized for specific tasks (e.g., straight stitch only)
- Larger, more robust components

Popular Brands: Juki, Brother, Consew

Common Issues:
- Motor burnout
- Hook and bobbin case wear
- Oil leakage

Troubleshooting Tip: For industrial machines, establish a strict oiling schedule and stick to it to prevent premature wear and potential breakdowns.

6. Vintage Sewing Machines

Older machines, often prized for their durability and all-metal construction.

Characteristics:
- Simple, mechanical designs
- Heavy, cast-iron bodies
- Often require specialized attachments

Popular Brands: Singer (pre-1970s), Necchi, Elna

Common Issues:
- Seized mechanisms due to old lubricant
- Electrical system degradation
- Finding replacement parts

Troubleshooting Tip: When restoring vintage machines, start by thoroughly cleaning all moving parts with a solvent to remove old, sticky lubricant before applying fresh oil.

Brand-Specific Considerations

While the general principles of sewing machine repair apply across brands, each manufacturer has its quirks and specialties. Here's a brief overview of some major brands:

1. Singer
   - Oldest and most recognizable brand
   - Vast range of models from vintage to modern
   - Excellent documentation available for most models

2. Brother
   - Known for user-friendly, feature-rich computerized machines
   - Often incorporates innovative technology
   - Can be challenging to repair due to proprietary parts

3. Janome
   - Reputation for reliability and consistent stitch quality
   - Offers a wide range of machines for different skill levels
   - Generally easy to maintain and repair

4. Bernina
   - High-end machines with precision engineering
   - Known for excellent stitch quality and durability
   - Can be complex to repair due to sophisticated mechanisms

5. Juki
- Popular for both domestic and industrial machines
- Known for high-speed, heavy-duty performance
- Industrial models require specialized knowledge to repair

Troubleshooting Tip: Always refer to the specific model's manual when possible, as procedures can vary significantly even within the same brand.

General Repair Approach

Regardless of the type or brand of machine you're working on, follow these steps:

1. Identify the machine type and model
2. Consult the manual or find online resources specific to that model
3. Perform a visual inspection and basic maintenance (cleaning, oiling)
4. Test the machine to identify the specific issue
5. Research common problems for that machine type/brand
6. Start with the simplest potential fixes before moving to more complex repairs

Troubleshooting Tip: Create a checklist for each machine type you commonly work on, listing the most frequent issues and their solutions. This can save time and ensure you don't overlook simple fixes.

By understanding the different types of sewing machines and the unique characteristics of various brands, you'll be better equipped to approach repairs with confidence. Remember, each machine has its own personality, and with experience, you'll develop an intuition for diagnosing and fixing issues across a wide range of models.

In the next section, we'll explore the anatomy of sewing machines in detail, giving you a comprehensive understanding of how these fascinating devices work. Are you ready to peek under the hood and discover the intricate mechanisms that bring fabric and thread together?

Chapter 2
Sewing Machine Basics
Anatomy of a Sewing Machine

Understanding the anatomy of a sewing machine is crucial for effective repair and maintenance. Let's break down the major components, their functions, and common issues associated with each part.

1. The Head

The head is the main body of the sewing machine, housing most of the working parts.

Key Components:
- Spool Pin: Holds the thread spool
- Thread Guide: Directs thread from the spool to the tension assembly
- Tension Assembly: Controls the tightness of the upper thread
- Take-Up Lever: Pulls thread from the spool and controls thread slack

Troubleshooting Tip: If you're experiencing loose stitches on the top of the fabric, check the tension assembly for proper threading and adjust as needed.

2. The Needle Assembly

This crucial component is responsible for carrying the upper thread through the fabric.

Key Components:
• Needle Bar: Holds the needle and moves it up and down
• Needle Clamp: Secures the needle in place
• Presser Foot: Holds the fabric in place during sewing
• Presser Foot Lifter: Raises and lowers the presser foot

Troubleshooting Tip: If you're breaking needles frequently, ensure the needle is inserted correctly (flat side to the back) and is the right size for your fabric.

3. The Bobbin System

The bobbin system provides the lower thread, which interlocks with the upper thread to form stitches.

Key Components:
• Bobbin: Holds the lower thread
• Bobbin Case: Houses the bobbin
• Hook: Catches the upper thread to form a loop around the bobbin thread
• Feed Dogs: Move the fabric through the machine

Troubleshooting Tip: If your machine is skipping stitches, check that the bobbin is inserted correctly and the bobbin area is free of lint and thread debris.

4. The Feed System

This system is responsible for moving the fabric through the machine as you sew.

Key Components:
• Feed Dogs: Textured metal bars that grip and move the fabric

• Throat Plate: Metal plate with holes for the needle and feed dogs
• Presser Foot: Works with the feed dogs to guide the fabric

Troubleshooting Tip: If your fabric isn't feeding evenly, check that the presser foot is lowered and the feed dogs are engaged (some machines allow you to drop the feed dogs for free-motion sewing).

5. The Motor and Drive System

These components power the machine and transfer motion to the various moving parts.

Key Components:
• Motor: Provides power to the machine
• Drive Belt or Shaft: Transfers power from the motor to the main shaft
• Hand Wheel: Allows manual control of the needle position
• Clutch (on some models): Disengages the needle for bobbin winding

Troubleshooting Tip: If your machine is running sluggishly, check the drive belt for wear or slippage. A worn belt can significantly reduce power transfer from the motor.

6. The Stitch Selection Mechanism

This system allows you to choose different stitch types and adjust stitch length and width.

Key Components:
• Stitch Selector: Dial or buttons to choose stitch patterns
• Length and Width Controls: Adjust stitch dimensions
• Reverse Lever or Button: Allows backward stitching

Troubleshooting Tip: If your machine isn't producing the selected stitch pattern, ensure that the width and length settings are appropriate for that stitch type.

7. The Lighting System

Proper illumination is crucial for accurate sewing.

Key Components:
- Sewing Light: Illuminates the sewing area
- Light Switch: Controls the sewing light

Troubleshooting Tip: If your sewing light isn't working, check if the bulb needs replacing before assuming there's an electrical issue.

8. The Tension System

Proper tension is crucial for forming balanced stitches.

Key Components:
- Upper Tension Assembly: Controls upper thread tension
- Bobbin Tension: Controlled by a small screw on the bobbin case

Troubleshooting Tip: When adjusting tension, make small, incremental changes and test on scrap fabric. Remember, the lower the number, the looser the tension.

9. The Frame and Base

These provide structure and stability to the machine.

Key Components:
• Machine Body: Houses internal components
• Free Arm (on some models): Allows sewing of tubular items
• Carrying Handle: For portable machines

Troubleshooting Tip: If your machine is vibrating excessively during use, check that it's on a stable surface and that the rubber feet are intact.

10. Computerized Components (for modern machines)

These parts control the advanced features of computerized sewing machines.

Key Components:
• LCD Screen: Displays stitch selection and machine status
• Control Panel: Buttons or touchscreen for selecting functions
• Microprocessor: Controls the machine's computerized functions

Troubleshooting Tip: If your computerized machine is acting erratically, try resetting it to factory settings (consult your manual for instructions).

Understanding Interconnections

While it's important to know individual components, understanding how they work together is crucial for effective troubleshooting:

- The motor drives the main shaft, which in turn moves the needle bar, feed dogs, and hook.
- The take-up lever, tension assembly, and hook must be precisely timed to form proper stitches.
- The bobbin system and upper threading must work in harmony to create balanced stitches.

Advanced Troubleshooting Tip: When dealing with stitch quality issues, consider the entire stitch formation process. A problem with uneven stitches could be caused by the tension assembly, bobbin tension, or even the hook timing.

Maintenance Insights

Regular maintenance of these components is key to your machine's longevity:

- Clean the feed dogs and hook area after every few projects.
- Oil the machine as recommended in the manual, typically focusing on moving parts in the head.
- Regularly check and tighten screws, especially those on the needle clamp and presser foot.

By understanding the anatomy of your sewing machine, you'll be better equipped to diagnose issues, perform routine maintenance, and even tackle more complex repairs. Remember, each machine model may have slight variations, so always consult your specific machine's manual when in doubt.

In the next section, we'll explore how these components work together to create stitches, giving you a deeper understanding of the sewing process. Are you ready to unravel the mysteries of stitch formation?

How Sewing Machines Work

Understanding the mechanics behind sewing machines is crucial for effective repair and maintenance. Let's break down the sewing process step-by-step, exploring how various components work together to create stitches.

The Basic Stitch Formation Process

1. Threading the Machine
  • Upper threading: Thread passes from the spool through guides, tension discs, take-up lever, and finally through the needle eye.
  • Lower threading: Bobbin is wound and placed in the bobbin case beneath the needle plate.

Troubleshooting Tip: Many issues stem from incorrect threading. Always rethread both upper and lower threads if you're experiencing unexpected problems.

2. Starting the Stitch
  • As you press the foot pedal, the motor activates, turning the main drive shaft.
  • This rotation is transferred to various mechanisms through a series of gears, cams, and levers.

3. Needle Descent
  • The needle bar moves downward, pushing the threaded needle through the fabric.
  • As the needle reaches its lowest point, it creates a small loop of thread on the underside of the fabric.

Troubleshooting Tip: If the needle isn't penetrating the fabric fully, check for a bent needle or incorrect needle size for the fabric thickness.

4. Hook Rotation
- As the needle begins to rise, the rotating hook catches the loop of upper thread.
- The hook carries this loop around the bobbin case, wrapping it around the lower thread.

5. Loop Formation
- The hook continues to rotate, enlarging the loop of upper thread.
- This loop is guided around the bobbin case, encircling the lower thread.

Troubleshooting Tip: Timing issues between the needle and hook can cause skipped stitches. If this occurs consistently, the hook timing may need adjustment.

6. Thread Take-Up
- As the needle reaches its highest point, the take-up lever pulls the excess thread, tightening the loop around the bobbin thread.
- This action pulls both threads tight, forming a complete stitch.

7. Fabric Feed
- The feed dogs rise above the throat plate, gripping the fabric.
- They then move backward, pulling the fabric along for the next stitch.
- The presser foot applies pressure from above, ensuring smooth fabric movement.

Troubleshooting Tip: If fabric isn't feeding evenly, check for lint buildup around the feed dogs and ensure the presser foot is applying adequate pressure.

8. Tension Balance
 • Throughout this process, the tension system ensures that the upper and lower threads interlock in the middle of the fabric layers.
 • Proper tension is crucial for creating balanced, strong stitches.

Troubleshooting Tip: Uneven stitches often indicate tension issues. Adjust upper tension first, then lower if needed.

Advanced Mechanics: Different Stitch Types

While the basic lockstitch is the foundation of machine sewing, understanding how machines create various stitch types can aid in troubleshooting more complex issues.

1. Straight Stitch
 • The most basic stitch type.
 • Needle moves only up and down; feed dogs move fabric forward.

2. Zigzag Stitch
 • Needle bar moves side to side as well as up and down.
 • Controlled by a cam mechanism or electronic signals in computerized machines.

Troubleshooting Tip: If zigzag width is inconsistent, check for worn cams in mechanical machines or calibration issues in computerized models.

3. Buttonhole Stitch
 • Combines straight and zigzag stitches.
 • Often uses a specialized presser foot and multi-step process.

• In advanced machines, this is automated through a series of preset movements.

4. Stretch Stitches
 • Involve forward and backward fabric movement.
 • Achieved through specialized feed dog movements or needle bar patterns.

Troubleshooting Tip: Stretch stitches not forming correctly often indicate timing issues between needle movement and feed dog action.

5. Decorative Stitches
 • Created through complex combinations of needle and fabric movements.
 • In computerized machines, these are controlled by precise electronic signals to various components.

The Role of Key Components

Understanding how specific components contribute to stitch formation can help in diagnosing issues:

1. Bobbin and Bobbin Case
 • Supplies lower thread.
 • Bobbin case often has a small tension spring to control lower thread tension.

Troubleshooting Tip: A poorly wound bobbin can cause inconsistent lower tension. Always wind bobbins evenly and at consistent speed.

2. Feed Dogs
  • Control fabric movement.
  • Different stitch lengths are achieved by varying the distance of feed dog movement.

3. Needle
  • Carries upper thread through fabric.
  • Different needle types and sizes are crucial for different fabrics and threads.

Troubleshooting Tip: Using the wrong needle type can cause skipped stitches, fabric damage, or thread breakage.

4. Hook Assembly
  • Catches upper thread loop to form stitch.
  • Timing between hook and needle is crucial for proper stitch formation.

5. Tension Discs
  • Control upper thread tension.
  • Must release tension when presser foot is raised to allow for easy fabric positioning.

Troubleshooting Tip: Check that tension discs open fully when presser foot is raised. If not, the mechanism may need cleaning or repair.

6. Take-Up Lever
  • Controls thread slack and tightens each stitch.
  • Its movement must be precisely timed with needle and hook movement.

Advanced Troubleshooting: Timing and Synchronization

Many complex sewing issues stem from timing problems between various components:

1. Needle-Hook Timing
 • The hook should catch the thread loop when the needle is on its upward stroke, about 3mm above its lowest point.
 • Incorrect timing can cause skipped stitches or thread breakage.

2. Feed Dog-Needle Bar Synchronization
 • Feed dogs should be below the throat plate when the needle enters the fabric.
 • Poor synchronization can cause fabric damage or uneven stitches.

3. Take-Up Lever Timing
 • Should reach its highest point as the needle reaches its highest point.
 • Incorrect timing can result in loose upper stitches or thread breakage.

Troubleshooting Tip: If you suspect timing issues, consult your machine's service manual for specific timing marks and adjustment procedures. This is often a job best left to experienced technicians.

Understanding how sewing machines work is key to effective troubleshooting and repair. By grasping these mechanics, you'll be better equipped to diagnose issues, perform maintenance, and even explain problems to professional technicians when necessary.

In our next section, we'll explore common sewing machine problems and their solutions, building on this foundational knowledge of sewing machine operation. Are you ready to put your new understanding to the test and tackle some real-world sewing machine issues?

Common Problems and Quick Fixes

Even the most reliable sewing machines can develop issues over time. Understanding these common problems and their solutions will empower you to keep your machine running smoothly.

1. Thread Bunching or Nesting

Problem: Threads forming a tangled mess on the underside of the fabric.

Causes:
- Incorrect upper threading
- Incorrect bobbin threading
- Unbalanced tension
- Dirty bobbin area

Quick Fixes:
a) Rethread the upper thread, ensuring it's seated correctly in the tension discs.
b) Check bobbin threading and tension.
c) Clean the bobbin area thoroughly.
d) Adjust upper tension if needed.

Troubleshooting Tip: Always start by rethreading both upper and lower threads. This simple step often resolves the issue without further intervention.

2. Skipped Stitches

Problem: The machine fails to form stitches at regular intervals.

Causes:
- Bent or dull needle
- Incorrect needle size for the fabric
- Improper threading
- Timing issues

Quick Fixes:

a) Replace the needle with a new, appropriate size and type for your fabric.
b) Rethread the machine, ensuring thread is properly seated in the tension discs.
c) Check that the needle is inserted correctly (flat side to the back).
d) Clean the hook race and check for any burrs or damage.

Troubleshooting Tip: If skipped stitches persist after trying these fixes, the machine may need professional timing adjustment.

3. Breaking Needles

Problem: Needles snapping frequently during sewing.

Causes:
- Pulling fabric while sewing
- Incorrect needle size or type
- Needle inserted incorrectly
- Misaligned needle plate

Quick Fixes:

a) Let the feed dogs move the fabric; guide it gently without pulling.
b) Ensure you're using the correct needle size and type for your fabric.

c) Check that the needle is inserted fully and correctly oriented.
d) Examine the needle plate for signs of damage or misalignment.

Troubleshooting Tip: If needles break consistently at the same point in your sewing, check for a burr on the needle plate or hook that might be catching the needle.

4. Uneven Stitches

Problem: Stitches vary in length or appear irregular.

Causes:
- Inconsistent fabric feeding
- Incorrect presser foot pressure
- Tension imbalance
- Worn or dirty feed dogs

Quick Fixes:
a) Clean the feed dogs and check for wear.
b) Adjust presser foot pressure if your machine allows.
c) Ensure you're using the correct presser foot for your fabric and stitch type.
d) Check and adjust thread tensions.

Troubleshooting Tip: Test sew on a double layer of medium-weight cotton. If stitches are even on this, the issue might be related to how you're handling more challenging fabrics.

5. Thread Breaking

Problem: Upper or lower thread snaps during sewing.

Causes:
- Poor quality or old thread
- Incorrect threading
- Burrs on thread path
- Excessive upper tension

Quick Fixes:
a) Replace with high-quality, new thread.
b) Rethread the machine, checking for proper path through all guides.
c) Check for and smooth any burrs on the thread path, needle plate, or bobbin case.
d) Reduce upper tension slightly.

Troubleshooting Tip: If thread breaks when sewing at high speeds, try sewing more slowly. Some threads, especially metallic ones, require slower sewing speeds.

6. Machine Running Slowly or Sluggishly

Problem: Machine doesn't maintain speed or seems to lack power.

Causes:
- Lint and debris buildup
- Need for oiling
- Worn or loose drive belt
- Electrical issues

Quick Fixes:
a) Perform a thorough cleaning of all accessible parts.
b) Oil the machine according to the manual's instructions.
c) Check the drive belt for wear and proper tension.
d) Ensure the machine is properly plugged in and the outlet is functioning.

Troubleshooting Tip: If your machine has multiple speed settings, test at different speeds to isolate whether the issue is mechanical or electronic.

7. Fabric Not Feeding Properly

Problem: Fabric doesn't move smoothly under the presser foot.

Causes:
- Feed dogs lowered or clogged with lint
- Incorrect presser foot pressure
- Stitch length set to zero
- Worn feed dogs

Quick Fixes:
a) Ensure feed dogs are raised and clean of lint.
b) Adjust presser foot pressure for your fabric type.
c) Check that stitch length is set appropriately.
d) Examine feed dogs for wear and replace if necessary.

Troubleshooting Tip: For very light or very heavy fabrics, consider using a walking foot to assist with even feeding.

8. Loud Noises or Unusual Sounds

Problem: Machine makes knocking, clicking, or grinding noises.

Causes:
- Need for oiling
- Lint or thread jammed in hook race
- Bent needle hitting the needle plate
- Timing issues

Quick Fixes:
a) Oil the machine thoroughly according to the manual.
b) Clean the hook race area and check for thread jams.
c) Replace the needle and check for proper insertion.
d) Inspect for any visible obstructions in moving parts.

Troubleshooting Tip: Isolate the sound by manually turning the handwheel slowly. This can help pinpoint where the noise is coming from.

9. Buttonholes Not Forming Correctly

Problem: Uneven or improperly sized buttonholes.

Causes:
- Incorrect presser foot
- Unbalanced tension
- Inconsistent fabric feeding
- Buttonhole lever not lowered

Quick Fixes:
a) Ensure you're using the correct buttonhole foot.
b) Balance upper and lower tensions.
c) Use stabilizer under lightweight fabrics.
d) Double-check that the buttonhole lever is fully lowered.

Troubleshooting Tip: Always test buttonholes on scrap fabric identical to your project fabric, including any interfacing or stabilizers.

10. Needle Threading Difficulties

Problem: Unable to thread the needle easily.

Causes:
- Bent or dull needle
- Incorrect threading technique
- Poor lighting
- Eye strain

Quick Fixes:
a) Replace with a new, straight needle.
b) Ensure the presser foot is raised when threading to release tension discs.
c) Improve lighting in your sewing area.
d) Use a needle threader tool for assistance.

Troubleshooting Tip: If your machine has an automatic needle threader, ensure it's properly aligned and free of lint or thread debris.

General Troubleshooting Tips:

1. **Restart from Scratch:** When facing persistent issues, completely unthread the machine (top and bobbin), replace the needle, and rethread everything. This often resolves multiple problems.

2. **Use Quality Materials:** Always use high-quality thread and needles appropriate for your fabric. Poor quality materials can cause numerous issues.

3. **Regular Maintenance:** Clean and oil your machine regularly according to the manual's instructions. This prevents many common problems.

4. **Read the Manual:** Your machine's manual is an invaluable resource. Familiarize yourself with its troubleshooting section.

5. **Test on Scrap Fabric:** Before starting a project, always test your settings on scrap fabric identical to your project material.

6. **Keep a Maintenance Log:** Track when you clean, oil, and perform maintenance on your machine. This can help identify patterns in recurring issues.

7. **Know When to Seek Help:** While many problems can be solved at home, some issues require professional servicing. Don't hesitate to consult a technician for complex problems, especially those involving timing or electrical components.

By familiarizing yourself with these common problems and their solutions, you'll be well-equipped to handle most sewing machine issues that arise. Remember, patience and systematic troubleshooting are key to resolving sewing machine problems effectively.

In our next section, we'll delve into more specific repair techniques for different types of sewing machines. Are you ready to take your troubleshooting skills to the next level?

Chapter 3
Vintage Sewing Machine Repair
Identifying Vintage Models

Vintage sewing machines are not just tools; they're pieces of history. Identifying these models accurately is crucial for proper repair and maintenance. Let's explore the key aspects of identifying vintage sewing machines.

1. Age Classification

Generally, sewing machines are considered vintage if they're over 25 years old, but true antiques are 100+ years old.

- Pre-1900: Truly antique, often featuring ornate decals and cast iron construction
- 1900-1960: Classic vintage era, including many iconic models
- 1960-1990: Late vintage, transitioning to more modern features

Troubleshooting Tip: When working on pre-1960 machines, be prepared for non-standardized parts and imperial measurements rather than metric.

2. Brand Identification

Major vintage brands include:

- Singer: The most common vintage brand
- White
- Kenmore
- Necchi
- Pfaff

• Elna

Look for brand names on the machine body, often in raised lettering or decals.

3. Model Numbers and Serials

• Singer: Look for the serial number on the base or bed of the machine. Use Singer's online database to date the machine precisely.
• Other brands: Serial numbers are usually on the bed, base, or inside the bobbin compartment.

Troubleshooting Tip: If you can't find a serial number, check under the machine or inside compartments. Some numbers may be hidden under years of grime.

4. Physical Characteristics

a) Body Style:
• Fiddle Base: Oldest style, pre-1900
• Bentwood Case: Popular in early 1900s
• Art Deco: Streamlined designs of the 1930s-1950s

b) Construction Material:
• Cast Iron: Most pre-1960s machines
• Aluminum: Became common in the 1960s
• Plastic Components: Started appearing in the late 1960s

c) Decals and Ornamentation:
• Elaborate Decals: Common on machines from 1850-1920
• Simpler Designs: Became prevalent in the 1930s-1950s
• Minimal Decoration: More common in later vintage models

Troubleshooting Tip: Be extremely careful when cleaning decorated surfaces. Use only gentle, non-abrasive cleaners to preserve delicate decals.

5. Mechanism Type

a) Treadle Machines:
- Foot-powered
- No electrical components
- Popular until the 1920s

b) Hand Crank Machines:
- Manually operated
- Portable
- Common in early 1900s

c) Electric Machines:
- Introduced in the 1920s
- Became standard by the 1950s

Troubleshooting Tip: When working on treadle or hand crank machines, focus on mechanical issues. These simple machines are often easier to repair than their electric counterparts.

6. Stitch Capabilities

- Straight Stitch Only: Most machines before the 1950s
- Zigzag Capability: Became common in the 1950s
- Multiple Stitch Patterns: Introduced in the 1960s

7. Bobbin Type

- Long Bobbin: Used in very early machines
- Round Bobbin: Became standard in the early 1900s
- Front-Loading vs. Top-Loading: Can indicate the era of the machine

Troubleshooting Tip: Always check the bobbin type before attempting repairs. Using the wrong bobbin can cause numerous issues.

8. Specific Features to Look For

a) Reverse Stitching:
- Introduced in the 1930s
- Became standard in the 1950s

b) Built-in Motor:
- Early electric machines had external motors
- Integrated motors became common in the 1950s

c) Cam System for Decorative Stitches:
- Introduced in the 1950s
- Indicates a later vintage model

9. Online Resources for Identification

- ISMACS (International Sewing Machine Collectors' Society): Offers extensive databases and information
- Fiddlebase.com: Specializes in dating and identifying Singer machines
- Victorian Sweatshop: Provides information on various vintage brands

Troubleshooting Tip: When using online resources, cross-reference information from multiple sources for accuracy.

10. Common Vintage Models and Their Characteristics

a) Singer 201:
- Known as the "Rolls Royce" of sewing machines
- Produced from the 1930s to 1950s
- Known for smooth, quiet operation

b) Singer Featherweight (221):
- Compact and portable
- Produced from 1933 to 1964
- Highly sought after by collectors

c) Elna Supermatic:
- First consumer machine with cam system for decorative stitches
- Introduced in 1952
- Known for its distinctive green color

d) Necchi Supernova:
- Innovative free-arm design
- Introduced in the 1950s
- Known for its wide range of built-in stitches

Troubleshooting Tip: Familiarize yourself with the quirks of popular models. For example, the Singer Featherweight is known for its sensitive tension system.

11. Assessing Condition

When identifying a vintage machine, also assess its condition:

- Rust or Corrosion: Common in machines stored in damp conditions
- Seized Mechanisms: Often due to old, hardened oil
- Electrical Issues: Common in early electric machines
- Missing Parts: Check against complete models to identify missing components

Troubleshooting Tip: Before starting any repair, thoroughly clean the machine and manually rotate the mechanism to identify any stiff or seized parts.

12. Documenting Your Findings

As you identify a vintage machine, document your findings:

- Take clear photos of the machine, including close-ups of identifying features
- Record all numbers and markings found on the machine
- Note any unique features or modifications

This documentation will be invaluable for research and when seeking advice from other enthusiasts or experts.

Identifying vintage sewing machines is both an art and a science. It requires attention to detail, research, and often a bit of detective work. By understanding the key features and characteristics of different eras and brands, you'll be better equipped to accurately identify vintage models, which is the crucial first step in proper repair and restoration.

In our next section, we'll explore the specific techniques for cleaning and restoring these beautiful machines. Are you ready to breathe new life into a piece of sewing history?

Cleaning and Oiling Antique Machines

Proper cleaning and oiling are essential for keeping antique sewing machines in working order. These processes require patience, attention to detail, and the right techniques to avoid damaging these delicate machines.

1. Safety First

Before you begin:
- Unplug the machine if it's electric
- Work in a well-ventilated area
- Wear gloves to protect your hands from oil and cleaning agents
- Use eye protection, especially when working with solvents

Troubleshooting Tip: If you encounter any electrical components, especially in machines from the 1920s-1950s, exercise extreme caution. Old wiring can be fragile and potentially dangerous.

2. Gathering Supplies

Essential supplies include:
- Soft lint-free cloths
- Old toothbrushes or small cleaning brushes
- Sewing machine oil
- Cleaning solvent (kerosene or mineral spirits)
- Screwdrivers (flathead and Phillips)
- Tweezers
- Cotton swabs
- Compressed air (optional)

Troubleshooting Tip: Never use WD-40 or household oils on sewing machines. These can gum up the mechanisms over time.

3. Initial Assessment

Before cleaning:
• Take photos of the machine from various angles for reference
• Make note of any areas with visible damage or excessive grime
• Identify all moving parts that will need oiling

4. Exterior Cleaning

Start with the machine's exterior:
a) Dust Removal:
 • Use a soft brush or compressed air to remove loose dust
 • Pay special attention to crevices and decorative elements

b) Surface Cleaning:
 • For painted surfaces, use a mild soap solution
 • For metal surfaces, use a cloth dampened with mineral spirits
 • Be extremely gentle with decals to avoid damaging them

c) Stubborn Grime:
 • Use a mixture of equal parts vinegar and water for tough dirt
 • For rust, gently buff with fine steel wool, being careful not to damage surrounding areas

Troubleshooting Tip: If you encounter sticky residue from old labels or tape, use a small amount of citrus-based adhesive remover, testing it on an inconspicuous area first.

5. Disassembly (if necessary)

Some cleaning may require partial disassembly:
• Refer to your machine's manual if available
• Take photos or draw diagrams as you disassemble to aid in reassembly
• Keep screws and small parts organized in labeled containers

Troubleshooting Tip: If a screw seems stuck, don't force it. Apply a penetrating oil and let it sit for a few hours before trying again.

6. Interior Cleaning

Focus on the machine's working parts:
a) Bobbin Area:
• Remove the bobbin case and clean thoroughly
• Use a brush to remove lint and debris from the bobbin housing
• Clean the feed dogs with a brush and cloth

b) Tension Assembly:
• Carefully clean between tension discs with a thin cloth or dental floss
• Avoid disassembling the tension unit unless absolutely necessary

c) Needle Bar and Presser Foot:
• Clean all visible parts with a cloth dampened with solvent
• Use cotton swabs for hard-to-reach areas

d) Handwheel and Belt:
  • Clean the handwheel, ensuring smooth rotation
  • If present, check the belt for wear and clean with a dry cloth

Troubleshooting Tip: If you encounter hardened oil or grease, especially in older machines, use a solvent-soaked cloth and let it sit on the area for a few minutes to soften the residue before wiping.

7. Oiling the Machine

Proper oiling is crucial for smooth operation:
a) Identify Oiling Points:
  • Consult the manual for specific oiling points
  • Generally, oil any point where metal moves against metal

b) Applying Oil:
  • Use only a few drops of oil at each point
  • Wipe away any excess oil to prevent attracting dust

c) Common Oiling Points:
  • Needle bar
  • Shuttle race
  • Handwheel mechanism
  • Feed dog mechanism

d) After Oiling:
  • Run the machine (without thread) to distribute the oil
  • Wipe away any oil that seeps out

Troubleshooting Tip: Over-oiling can be as problematic as under-oiling. If you notice oil dripping onto your fabric during sewing, you've likely over-oiled the machine.

8. Cleaning and Oiling Treadle Machines

For treadle machines, additional steps are necessary:
- Clean the treadle mechanism, removing any rust with fine steel wool
- Oil the treadle bearings and connections
- Check and clean the leather belt, replacing if necessary

9. Cleaning Wooden Parts

For machines with wooden bases or cases:
- Use a slightly damp cloth to clean, avoiding excessive moisture
- For stubborn dirt, use a mild wood cleaner
- Apply a small amount of wood polish to restore shine

10. Restoring Metal Surfaces

For machines with significant rust or tarnish:
- Use a metal polish appropriate for the type of metal (usually nickel or chrome for vintage machines)
- Apply with a soft cloth, working in small sections
- Buff to a shine with a clean cloth

Troubleshooting Tip: If you encounter deep rust pits, it may be best to leave them be. Aggressive cleaning can sometimes do more harm than good to the machine's integrity.

11. Final Steps

After cleaning and oiling:
- Reassemble any parts you removed, referring to your photos or diagrams
- Test the machine's movement by turning the handwheel

- If electric, carefully test the machine's operation
- Clean up any oil spots on your work surface

12. Ongoing Maintenance

To keep your antique machine in top condition:
- Clean the exterior regularly with a soft cloth
- Remove lint from the bobbin area after each use
- Oil the machine every 8-10 hours of use
- Store the machine covered to prevent dust accumulation

Troubleshooting Tip: Create a maintenance log to track when you've cleaned and oiled your machine. This can help you establish a routine and identify any recurring issues.

13. Special Considerations for Rare or Valuable Machines

For extremely old or valuable machines:
- Consider consulting a professional for cleaning and restoration
- Use only the gentlest cleaning methods to preserve original finishes
- Document the cleaning process for historical record-keeping

Cleaning and oiling antique sewing machines is a delicate process that requires patience and care. By following these steps, you can restore your machine to its former glory and ensure it continues to function smoothly for years to come. Remember, each machine is unique, so always be prepared to adapt these general guidelines to the specific needs of your antique sewing machine.

In our next section, we'll explore the challenging but rewarding process of restoring and replacing vintage parts. Are you ready to dive deeper into the world of antique sewing machine restoration?

Restoring and Replacing Vintage Parts

Restoring and replacing parts on vintage sewing machines requires a delicate balance of preserving historical integrity and ensuring functionality. This process can be both challenging and rewarding, often breathing new life into machines that have been silent for decades.

1. Assessing the Machine

Before starting any restoration:
• Thoroughly inspect the machine, identifying all damaged or missing parts
• Research your specific model to understand its original components
• Decide on your restoration goals: full historical accuracy or basic functionality

Troubleshooting Tip: Create a detailed inventory of parts, noting their condition. This will help you prioritize your restoration efforts and track down necessary replacements.

2. Sourcing Vintage Parts

Finding original parts can be challenging but is often crucial for authentic restoration:
• Online marketplaces (eBay, Etsy) specializing in vintage sewing machine parts
• Sewing machine repair shops that may have old stock
• Sewing machine collectors' forums and groups
• Salvage parts from non-functioning machines of the same model

Troubleshooting Tip: When buying parts online, always verify compatibility with your specific model and year. Even slight variations can cause issues.

3. Cleaning and Evaluating Parts

Before installation:
- Clean parts thoroughly using appropriate solvents
- Inspect for wear, cracks, or other damage
- Test moving parts for smooth operation

4. Common Parts to Restore or Replace

a) Belts:
- Often deteriorate over time
- Measure carefully for exact replacements
- Consider modern synthetic belts for improved durability

b) Bobbin Winders:
- Check for smooth operation
- Lubricate moving parts
- Replace rubber tire if hardened or cracked

c) Tension Assembly:
- Clean thoroughly, avoiding disassembly if possible
- Replace tension springs if stretched or weakened

d) Gears:
- Inspect for worn or broken teeth
- Clean and re-lubricate
- Consider professional help for gear replacement, as it often affects timing

Troubleshooting Tip: When dealing with gears, mark their position before removal. Incorrect reassembly can throw off the machine's timing.

5. Restoring Metal Parts

For rusted or corroded parts:
• Use a rust remover solution, following product instructions carefully
• For light rust, try soaking in white vinegar overnight
• Use fine steel wool or a brass brush for gentle abrasion
• Apply a protective coating after cleaning to prevent future rust

6. Repairing Wooden Components

For machines with wooden bases or cases:
• Fill cracks or holes with wood filler
• Sand gently to restore smooth surfaces
• Apply wood stain to match original color
• Finish with a protective clear coat

Troubleshooting Tip: When repairing wooden components, always work in a well-ventilated area and allow ample drying time between steps.

7. Electrical Components

For electric vintage machines:
• Inspect wiring for fraying or damage
• Replace old fabric-covered cords with modern, safe alternatives
• Consider having a professional check and update the wiring if you're unsure

8. Restoring Decorative Elements

Many vintage machines have beautiful decals or paintwork:
• Clean gently with a soft cloth and mild soap solution
• For badly damaged decals, consider professional restoration or custom reproductions
• Touch up paint chips with enamel paint, matching colors carefully

Troubleshooting Tip: When touching up paint, less is more. Start with tiny amounts and build up gradually to avoid noticeable differences.

9. Replacing Gaskets and Seals

• Often deteriorate over time, leading to oil leaks
• Measure carefully and cut new gaskets from appropriate materials
• Consider making a template from the old gasket before removal

10. Restoring the Finish

For machines with dulled or damaged finishes:
• Clean thoroughly with mineral spirits
• For painted surfaces, use automotive polishing compounds
• For bare metal, consider re-plating severely damaged areas

11. Lubricating Restored Parts

After restoration:
• Use proper sewing machine oil on all moving parts
• Apply sparingly to avoid attracting dust
• Run the machine to distribute oil evenly

12. Adjusting and Timing

After replacing parts, the machine may need adjusting:
- Check and adjust the timing between the needle and hook
- Ensure proper tension in the upper thread mechanism
- Adjust the feed dog height if necessary

Troubleshooting Tip: If you're unsure about adjusting timing, consult a professional. Incorrect timing can damage the machine or produce poor stitches.

13. Testing Restored Parts

- Slowly operate the machine by hand, checking for smooth movement
- Test each function individually before full operation
- Sew on scrap fabric to check stitch quality

14. Documenting the Restoration

- Take "before and after" photos of restored parts
- Keep a log of all replacements and repairs
- Save any original parts you've replaced, they may be valuable to collectors

15. Challenges and Considerations

- Balancing authenticity with functionality
- Dealing with discontinued parts
- Understanding the value impact of non-original replacements

Troubleshooting Tip: When faced with a choice between authenticity and functionality, consider your primary goal for the machine. Is it a display piece or will it be used regularly?

16. 3D Printing and Modern Solutions

For truly irreplaceable parts:
• Consider 3D printing replicas (especially useful for small, non-stress bearing parts)
• Explore modern materials that can be adapted to fit vintage machines

17. Professional Help

Know when to seek professional assistance:
• For valuable or rare machines
• When dealing with complex timing issues
• If you encounter electrical problems beyond your expertise

Restoring and replacing vintage sewing machine parts is a meticulous process that requires patience, research, and often creativity. Each machine presents unique challenges, but the reward of bringing a piece of history back to life is immeasurable. Remember, the goal is not always perfection, but rather preserving the machine's character while restoring its functionality.

By following these guidelines and approaching each restoration with care and respect for the machine's history, you can ensure that these beautiful pieces of craftsmanship continue to delight and serve for many more years to come.

In our next section, we'll explore the nuances of working with more modern domestic sewing machines. Are you ready to bridge the gap between vintage charm and contemporary convenience?

Chapter 4
Modern Domestic Sewing Machine Maintenance
Troubleshooting Electronic Sewing Machines

Electronic sewing machines offer a wide range of features and capabilities, but they also come with their own set of challenges when it comes to maintenance and repair. Understanding how to troubleshoot these machines can save you time, money, and frustration.

1. Understanding Electronic Sewing Machines

Modern electronic sewing machines typically feature:
- Computerized control systems
- LCD screens
- Multiple stitch options controlled by electronic selection
- Automated features like thread cutters and needle threaders

Troubleshooting Tip: Always start by consulting your machine's manual. Many modern machines have self-diagnostic features that can point you in the right direction.

2. Common Electronic Issues

a) Display Problems:
- Blank or flickering LCD screen
- Error messages
- Unresponsive touch controls

b) Stitch Selection Issues:
- Machine not producing selected stitch
- Unexpected changes in stitch patterns

c) Motor Problems:
- Machine won't start
- Intermittent operation

d) Sensor Malfunctions:
- Thread break sensors not working
- Bobbin sensors giving false readings

3. Basic Troubleshooting Steps

Before diving into specific issues, always start with these steps:
- Power cycle the machine (turn off, unplug, wait 1 minute, plug in, turn on)
- Check for and install any available firmware updates
- Ensure the machine is clean and properly oiled
- Verify that thread and bobbin are correctly installed

Troubleshooting Tip: Keep a log of any error codes or unusual behaviors. This can help identify patterns and assist technicians if professional service is needed.

4. Dealing with Display Issues

a) Blank Screen:
- Check power connection and cord integrity
- Verify brightness settings haven't been accidentally lowered
- Look for physical damage to the screen

b) Flickering or Distorted Display:
• Check for loose connections inside the machine (may require professional help)
• Update firmware if available

c) Unresponsive Touch Controls:
• Clean the screen gently with a slightly damp, lint-free cloth
• Recalibrate the touch screen if your model allows

Troubleshooting Tip: If the screen is physically damaged, it's often more cost-effective to replace the entire display unit rather than attempting a repair.

5. Resolving Stitch Selection Problems

a) Machine Not Producing Selected Stitch:
• Verify that the correct presser foot is installed for the chosen stitch
• Check that the needle position is appropriate for the stitch
• Ensure the feed dogs are in the correct position (raised or lowered)

b) Unexpected Stitch Changes:
• Look for stuck buttons on the control panel
• Check for software glitches by resetting the machine
• Verify that no custom stitch programs are accidentally activated

Troubleshooting Tip: Some machines have a "reset to default" option. Use this if you suspect settings have been inadvertently changed.

6. Addressing Motor Issues

a) Machine Won't Start:
 • Check the foot pedal connection
 • Verify that the bobbin winder hasn't been accidentally engaged
 • Ensure the presser foot is lowered

b) Intermittent Operation:
 • Look for loose internal connections (may require professional inspection)
 • Check for obstructions in the motor's moving parts
 • Verify that the power source is stable and providing sufficient voltage

Troubleshooting Tip: If you suspect power supply issues, try plugging the machine into a different outlet, preferably on a different circuit.

7. Sensor Malfunctions

a) Thread Break Sensors:
 • Clean the sensor area thoroughly
 • Check for thread debris caught in the sensor
 • Verify that the thread path is correct and tension is appropriate

b) Bobbin Sensors:
 • Ensure the bobbin is inserted correctly
 • Clean the bobbin area and sensor
 • Try a different bobbin to rule out issues with a specific bobbin

Troubleshooting Tip: Some machines allow you to disable certain sensors. While this isn't a long-term solution, it can help you determine if a sensor is the root cause of an issue.

8. Software and Firmware Issues

• Regularly check for and install firmware updates
• If issues occur after an update, look for a way to revert to a previous version
• For machines with advanced features, ensure any connected devices (like computers or tablets) have compatible software versions

9. Dealing with Error Codes

• Keep your manual handy for quick reference to error code meanings
• Some common error codes and their general meanings:
 - E1 or similar: Often indicates a power or motor issue
 - E2 or similar: May point to a problem with the foot pedal
 - E3 or similar: Could indicate a thread or bobbin sensor issue

Troubleshooting Tip: Search online forums specific to your machine model. Other users may have found solutions to common error codes.

10. Advanced Troubleshooting Techniques

a) Circuit Board Inspection:
• Look for visible signs of damage or burning on circuit boards
• Check for loose connections or detached components
• Be cautious – circuit board repair often requires specialized skills

b) Voltage Testing:
  • Use a multimeter to check for correct voltage at power input
  • Test continuity in foot pedal and power cords

c) Memory Reset:
  • Some machines have a hard reset option that clears all stored memory
  • Use with caution, as this will erase all saved settings and custom stitches

11. When to Seek Professional Help

Consider professional servicing if:
• You encounter repeated error codes that you can't resolve
• There are unusual noises or burning smells
• You notice visible damage to electronic components
• The machine is under warranty (to avoid voiding it)

Troubleshooting Tip: When taking your machine for professional service, provide a detailed description of the issues, including any error codes and the steps you've already taken to troubleshoot.

12. Preventive Maintenance for Electronic Machines

• Keep the machine clean, especially around sensors and electronic components
• Use a surge protector to safeguard against power fluctuations
• Regularly update firmware and software
• Avoid exposing the machine to extreme temperatures or humidity

13. Understanding Limitations

Remember that while many issues can be resolved at home, electronic sewing machines are complex devices. Some problems may require specialized tools or knowledge to address safely and effectively.

By following these troubleshooting steps and maintaining your electronic sewing machine properly, you can ensure it continues to perform at its best. Remember, patience and methodical problem-solving are key when dealing with electronic issues. Don't hesitate to consult your manual, online resources, or professional technicians when needed.

In our next section, we'll explore the specific challenges of repairing computerized sewing machines, delving even deeper into the world of high-tech sewing. Are you ready to unlock the full potential of your modern sewing machine?

Repairing Computerized Models

Computerized sewing machines represent the cutting edge of sewing technology, offering a wide array of features and capabilities. However, their complexity can make repairs challenging. Let's explore how to approach these high-tech machines.

1. Understanding Computerized Sewing Machines

Key components of computerized models include:
- Microprocessor or CPU
- Memory modules
- LCD or LED display
- Touch-sensitive controls
- Multiple sensors
- Stepper motors for precise control

Troubleshooting Tip: Before attempting any repairs, always check your machine's warranty status. Unauthorized repairs may void the warranty.

2. Common Issues in Computerized Models

a) Software Glitches:
- Unexpected error messages
- Freezing or unresponsive controls
- Incorrect stitch selection

b) Hardware Malfunctions:
- Display issues (blank, flickering, or distorted screen)
- Unresponsive buttons or touch controls
- Sensor failures (thread, needle position, fabric thickness)

c) Motor and Movement Problems:
- Erratic needle or feed dog movement
- Inconsistent speed control

d) Connectivity Issues:
- Problems with USB ports or Wi-Fi connections
- Difficulties updating firmware

3. Diagnostic Approach

Before diving into repairs:
- Run any built-in diagnostic programs your machine offers
- Check for and note down any error codes
- Update the machine's firmware to the latest version
- Perform a factory reset if possible (note: this will erase all custom settings)

Troubleshooting Tip: Keep a detailed log of all issues, error codes, and steps taken. This can be invaluable if you need to consult a professional later.

4. Software-Related Repairs

a) Addressing Error Messages:
- Consult the manual for specific error code meanings
- Follow recommended steps for each error code
- If persistent, try a complete power cycle (unplug for 5 minutes)

b) Dealing with Freezes or Unresponsiveness:
- Perform a soft reset (usually a specific button combination)
- If unsuccessful, try a hard reset (refer to manual for procedure)

c) Firmware Updates:
• Always download firmware from the official manufacturer's website
• Follow update instructions precisely
• Ensure stable power during the update process

Troubleshooting Tip: If a firmware update fails, don't panic. Many machines have a recovery mode. Consult your manual or the manufacturer's support for guidance.

5. Hardware Troubleshooting

a) Display Issues:
• Check connections between the display and main board
• Look for physical damage to the screen
• If touchscreen, recalibrate if your model allows

b) Unresponsive Controls:
• Clean the control panel gently with isopropyl alcohol
• Check for loose connections on the control board
• Look for signs of liquid damage or corrosion

c) Sensor Problems:
• Clean all sensor areas thoroughly
• Check for obstructions or thread debris
• Verify sensor connections to the main board

Troubleshooting Tip: When dealing with electronic components, always work in a static-free environment. Use an anti-static wrist strap if possible.

6. Motor and Movement Repairs

a) Erratic Needle Movement:
- Check for obstructions in the needle bar assembly
- Verify connections to the stepper motor
- Look for worn or damaged gears

b) Feed Dog Issues:
- Ensure feed dogs are clean and unobstructed
- Check the feed dog height adjustment
- Verify connections to the feed motor

c) Speed Control Problems:
- Clean the foot pedal internally
- Check the pedal's connection to the machine
- Verify the integrity of the speed control circuit

Troubleshooting Tip: When dealing with motor issues, listen for unusual noises. Grinding or clicking sounds can indicate mechanical problems, while humming or buzzing might suggest electrical issues.

7. Connectivity Repairs

a) USB Port Issues:
- Check for physical damage to the port
- Try multiple USB devices to isolate the problem
- Verify internal connections to the main board

b) Wi-Fi Connection Problems:
- Update the machine's Wi-Fi firmware if applicable
- Check compatibility with your router's settings
- Verify the antenna connection inside the machine

8. Advanced Repair Techniques

a) Circuit Board Inspection:
 • Look for burned components or swollen capacitors
 • Check for loose or corroded solder joints
 • Measure voltages at key points (refer to service manual)

b) Stepper Motor Testing:
 • Use a multimeter to check motor windings for continuity
 • Verify proper voltage supply to the motor
 • Check for smooth manual rotation (when unpowered)

c) Sensor Calibration:
 • Some models allow manual calibration of sensors
 • Follow manufacturer's procedures carefully
 • Use test fabric or tools provided by the manufacturer

Troubleshooting Tip: For advanced repairs, consider investing in a service manual for your specific model. These often contain detailed schematics and test procedures.

9. When to Seek Professional Help

Consider professional servicing if:
 • You encounter persistent software issues after attempted resets and updates
 • There's visible damage to circuit boards or other internal components
 • You're uncomfortable working with sensitive electronic components
 • The machine is valuable or still under warranty

10. Preventive Maintenance for Computerized Machines

• Keep the machine clean, especially around sensors and vents
• Use a surge protector to guard against power fluctuations
• Regularly update firmware and software
• Backup custom stitches and settings periodically
• Use the machine regularly to prevent issues from disuse

Troubleshooting Tip: Create a maintenance schedule reminder on your phone or calendar to ensure regular upkeep of your computerized machine.

11. Understanding Limitations and Future-Proofing

• Be aware that some components may become obsolete over time
• Consider the long-term availability of software updates
• Keep original packaging and accessories for potential future service needs

Repairing computerized sewing machines requires a blend of technical knowledge, patience, and sometimes specialized tools. While many issues can be resolved with careful troubleshooting, it's important to recognize when a problem is beyond your skill level. Remember, the goal is to get your machine running smoothly without causing additional damage.

By following these guidelines and approaching repairs methodically, you can often resolve issues with your computerized sewing machine and keep it in optimal condition for years to come.

In our next section, we'll explore the unique challenges of maintaining plastic components in modern sewing machines. Are you ready to learn how to keep every part of your high-tech sewing companion in top shape?

Maintaining Plastic Components

Modern sewing machines often incorporate numerous plastic components, which can present unique challenges in terms of maintenance and repair. Understanding how to properly care for these parts is crucial for extending the life of your machine and ensuring optimal performance.

1. Understanding Plastic Components in Sewing Machines

Common plastic parts include:
- Outer casing and body panels
- Bobbin cases and covers
- Thread guides and tension discs
- Gear components
- Buttons and dials
- Presser feet
- Spool pins

Troubleshooting Tip: Keep a record of all plastic components in your machine. This can help you quickly identify any parts that may need attention or replacement over time.

2. Common Issues with Plastic Components

a) Cracking or Breaking:
- Often due to impact, stress, or age
- Can compromise the machine's functionality and appearance

b) Warping:
- Usually caused by heat exposure or chemical interaction
- May affect precision components like bobbin cases serious damage.

c) Discoloration:
- Typically results from UV exposure or chemical reactions
- Primarily an aesthetic issue but can indicate material degradation

d) Wear and Tear:
- Friction points can develop rough surfaces or grooves
- May affect thread movement or part interactions

3. Preventive Maintenance for Plastic Parts

a) Regular Cleaning:
- Use a soft, lint-free cloth slightly dampened with water
- For stubborn dirt, use a mild soap solution (avoid harsh chemicals)
- Dry thoroughly after cleaning

b) Proper Storage:
- Keep the machine covered when not in use
- Store in a cool, dry place away from direct sunlight
- Use a dehumidifier in high-humidity environments

c) Gentle Handling:
- Avoid applying excessive force to plastic components
- Use the machine as intended, following the manufacturer's guidelines

Troubleshooting Tip: Develop a routine maintenance schedule, marking your calendar for regular checks and cleaning of plastic components.

4. Lubricating Plastic Parts

• Generally, plastic components should not be oiled
• If lubrication is necessary, use only products specifically recommended by the manufacturer
• Apply lubricants sparingly to avoid attracting dust and debris

5. Dealing with Static Electricity

Static can cause thread and fabric to stick to plastic surfaces:
• Use anti-static sprays designed for sewing machines
• Increase humidity in your sewing area
• Touch a grounded metal object before using the machine

Troubleshooting Tip: If static is a persistent problem, consider using a humidifier in your sewing room, especially during dry seasons.

6. Repairing Cracked or Broken Plastic

For minor cracks or breaks:
• Clean the area thoroughly
• Use a plastic-compatible adhesive (e.g., epoxy or super glue)
• Apply in thin layers, allowing each to dry completely
• Sand gently if needed for a smooth finish

For major damage:
• Consider replacing the part entirely
• Contact the manufacturer for original replacement parts

Troubleshooting Tip: Before attempting to repair a broken plastic part, assess whether the repair might affect the machine's operation. Safety and functionality should always be prioritized over aesthetics.

7. Addressing Warped Plastic Components

• Identify the cause of warping (heat, stress, chemical exposure)
• For slight warping, try gently reshaping the part while it's at room temperature
• For severe warping, replacement is often the best option

8. Maintaining Plastic Gears

• Inspect regularly for signs of wear or cracking
• Keep clean and free of thread debris
• If replacement is necessary, ensure you use the correct part for your model

Troubleshooting Tip: If you notice unusual noises or resistance when operating your machine, inspect the plastic gears immediately. Early detection of wear can prevent more

9. Caring for Plastic Presser Feet

• Clean after each use, especially when working with fusible materials
• Store properly to avoid warping or cracking
• Replace if worn or damaged, as they're crucial for proper stitch formation

10. Maintaining Plastic Bobbin Cases

• Clean thoroughly after each project
• Check for wear, especially around the tension spring
• Replace if you notice inconsistent lower thread tension

11. Protecting Against UV Damage

• Store the machine away from direct sunlight
• Use UV-resistant covers when the machine is not in use
• Consider applying a UV-resistant coating to exposed plastic parts

Troubleshooting Tip: If you notice discoloration on plastic parts, it may indicate UV damage. Assess these areas for potential weakening of the material.

12. Chemical Interactions to Avoid

• Keep solvents, alcohol-based cleaners, and harsh detergents away from plastic components
• Be cautious with fabric treatments and sprays near the machine
• If using adhesives, ensure they're compatible with the type of plastic in your machine

13. Dealing with Squeaky Plastic Parts

• Identify the source of the noise
• Clean the area thoroughly
• If persistent, apply a small amount of silicone-based lubricant (if recommended by the manufacturer)

14. When to Replace Rather Than Repair

Consider replacement when:
• The damage affects the functionality of the machine
• Repair attempts have been unsuccessful
• The cost of replacement is comparable to repeated repair attempts

Troubleshooting Tip: Keep a small fund for potential part replacements. This can help you make objective decisions about when to replace versus repair.

15. Sourcing Replacement Plastic Parts

• Always try to obtain original parts from the manufacturer first
• For discontinued models, look for new old stock (NOS) parts
• Consider 3D printing for hard-to-find components, ensuring precise measurements

16. Future-Proofing Your Machine

• Stay informed about your model's long-term support and part availability
• Consider machines with metal alternatives for crucial components when purchasing
• Keep your machine's manual and any spare parts in a safe place

Maintaining the plastic components of your sewing machine is crucial for its longevity and performance. By following these guidelines, you can prevent many common issues and address problems effectively when they do arise. Remember, gentle, consistent care is key to keeping your machine's plastic parts in top condition.

In our next section, we'll explore the world of industrial sewing machine repair, where the scale and complexity of machines present a whole new set of challenges. Are you ready to dive into the realm of heavy-duty sewing equipment?

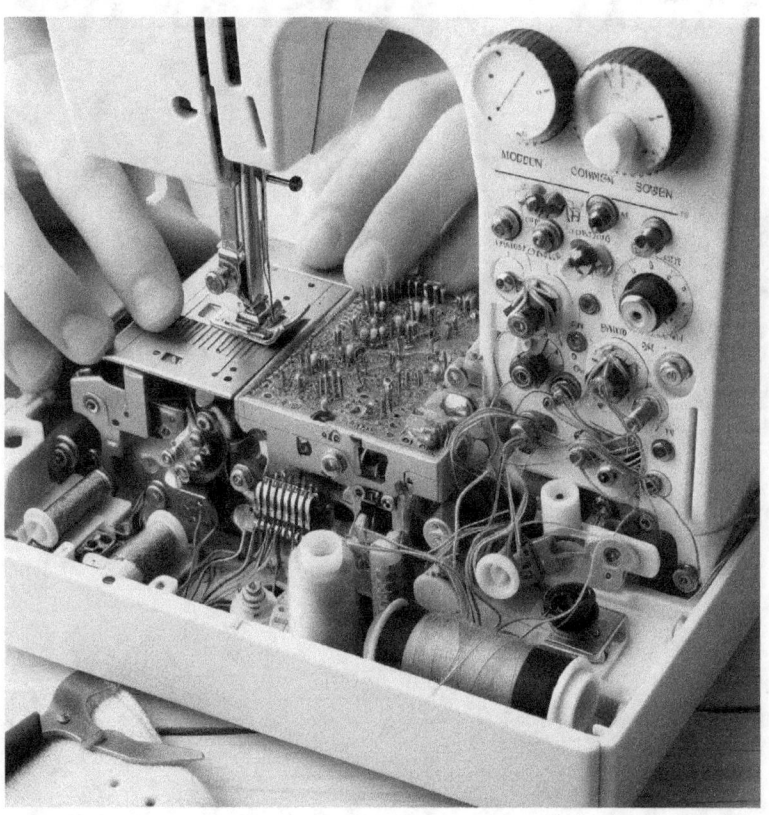

Chapter 5
Industrial Sewing Machine Repair
Understanding Industrial Machine Mechanics

Industrial sewing machines are built for high-volume, continuous operation and are significantly different from domestic models in terms of construction, power, and mechanics. Understanding these differences is crucial for effective repair and maintenance.

1. Characteristics of Industrial Sewing Machines

- Heavier construction with cast iron bodies
- Larger, more powerful motors (typically 1/2 to 1 horsepower)
- Higher operating speeds (up to 5000 stitches per minute or more)
- Specialized for specific tasks (e.g., straight stitch only, buttonholes, serging)
- Separate motor and light, often mounted underneath the table

Troubleshooting Tip: Always check the machine's specifications before beginning any repair. Industrial machines can vary greatly based on their intended use.

2. Key Components of Industrial Machines

a) Motor and Power Transmission:
- External motor connected via a belt or direct drive
- Clutch mechanism for engaging/disengaging power
- Speed reducer for precise control

b) Needle Bar and Hook Assembly:
- Larger, more robust components than domestic machines
- Often with specialized lubrication systems

c) Feed Mechanisms:
- Various types: drop feed, needle feed, walking foot, puller feed
- Designed for handling heavier materials

d) Tension Systems:
- More complex with multiple tension regulators
- Often include separate tensions for different threads (e.g., loopers in overlock machines)

e) Lubrication System:
- Many industrial machines have automatic or semi-automatic oiling systems
- Some require manual lubrication at specific points

Troubleshooting Tip: Familiarize yourself with the lubrication points and schedule for your specific machine. Proper lubrication is critical for industrial machines.

3. Understanding the Power Transmission System

- Belt Drive: Most common in older machines
  - Check belt tension and alignment regularly
  - Replace belts at the first sign of wear or fraying

- Direct Drive: More common in modern machines
  - Quieter operation and more energy-efficient
  - Requires less maintenance but can be more complex to repair

- Clutch and Brake Mechanism:
- Allows for instant stopping and starting
- Critical for safety and precision in high-speed operation

Troubleshooting Tip: If you notice delayed stopping or starting, inspect the clutch and brake mechanism immediately. Worn clutches can be a safety hazard.

4. Feed Mechanisms in Industrial Machines

- Drop Feed: Similar to domestic machines but more robust
- Needle Feed: Moves fabric in sync with needle movement
- Walking Foot: Upper and lower feed dogs move together
- Puller Feed: Additional feed mechanism behind the needle

Understanding these mechanisms is crucial for troubleshooting feeding issues.

5. Hook and Bobbin Systems

- Vertical Axis Hook: Common in lockstitch machines
- Horizontal Axis Hook: Used in some specialized machines
- Oscillating vs. Rotary Hooks: Affects timing and maintenance needs

Troubleshooting Tip: When dealing with timing issues, always check the relationship between the hook and needle. Even slight misalignments can cause significant problems.

6. Tension Systems in Industrial Machines

- Pre-tension: Initial thread control before main tension
- Main Tension: Primary tension control, often with numbered dials

- Check Spring: Controls thread loop formation
- Take-up Spring: Manages thread slack during stitch formation

7. Lubrication Systems

- Manual: Requires regular oiling at specified points
- Semi-automatic: Oil reservoirs that need periodic refilling
- Fully Automatic: Continuous lubrication during operation

Troubleshooting Tip: Over-oiling can be as problematic as under-oiling. Follow manufacturer guidelines precisely for lubrication.

8. Electrical Systems in Industrial Machines

- Motor: Usually more powerful and separate from the machine head
- Control Box: Houses speed control and other electronic functions
- Positioning Systems: For precise needle positioning (in advanced models)

9. Common Mechanical Issues and Troubleshooting

a) Timing Problems:
- Symptoms: Skipped stitches, thread breakage
- Check: Hook timing, needle bar height, take-up lever position

b) Feed Issues:
- Symptoms: Uneven stitch length, fabric damage
- Check: Feed dog height and condition, presser foot pressure

c) Tension Inconsistencies:
• Symptoms: Loose or tight stitches, puckering
• Check: All tension components, including pre-tension and check springs

d) Noisy Operation:
• Symptoms: Unusual sounds during operation
• Check: Lubrication, loose parts, worn bearings or gears

Troubleshooting Tip: Develop a systematic approach to diagnosing problems. Start with the simplest possible cause and work your way to more complex issues.

10. Preventive Maintenance for Industrial Machines

• Daily: Clean lint and debris, check oil levels
• Weekly: Inspect belts, check for loose screws
• Monthly: Thorough cleaning, oil change if necessary
• Annually: Complete overhaul, replace worn parts

11. Safety Considerations

• Always disconnect power before performing maintenance
• Use proper lifting techniques when moving heavy machine heads
• Be cautious of high-speed moving parts during operation

12. Advanced Repair Techniques

• Gear Replacement: Often requires specialized tools and precise adjustment
• Motor Rewinding: Consider professional service for motor issues
• Needle Bar Alignment: Critical for proper stitch formation

Troubleshooting Tip: Keep a repair log for each machine, noting all maintenance and repairs. This history can be invaluable for diagnosing recurring issues.

13. Tools for Industrial Machine Repair

- Specialized screwdrivers and wrenches
- Timing gauges and alignment tools
- Stroboscope for high-speed timing checks
- Multimeter for electrical diagnostics

14. Resources for Learning

- Manufacturer service manuals
- Industrial sewing machine repair courses
- Online forums and communities dedicated to industrial machines

Understanding industrial sewing machine mechanics requires a blend of mechanical aptitude, electrical knowledge, and specialized training. These machines are built for performance and longevity, but they also demand proper maintenance and skilled repair when issues arise.

By familiarizing yourself with the unique aspects of industrial sewing machines, you'll be better equipped to maintain and repair these powerful tools, ensuring they continue to perform at their best in demanding production environments.

In our next section, we'll delve into the specific techniques for timing and adjusting heavy-duty machines, a critical skill for any industrial sewing machine technician. Are you ready to master the precision required for these high-performance machines?

Timing and Adjusting Heavy-Duty Machines

Proper timing and adjustment are critical for the optimal performance of heavy-duty sewing machines. These processes ensure that all components work in perfect synchronization, producing high-quality stitches at high speeds.

1. Understanding the Importance of Timing

Timing in sewing machines refers to the synchronized movement of the needle, hook, and feed mechanism. Proper timing ensures:
- Correct stitch formation
- Reduced thread breakage
- Prevention of skipped stitches
- Optimal machine performance and longevity

Troubleshooting Tip: If you're experiencing persistent stitch quality issues, timing problems are often the culprit. Always check timing before assuming part replacement is necessary.

2. Key Components Involved in Timing

a) Needle Bar:
- Controls the vertical movement of the needle

b) Hook (Rotary or Oscillating):
- Catches the thread loop to form stitches

c) Feed Dogs:
- Move the fabric through the machine

d) Take-up Lever:
• Manages thread tension and slack

3. Basic Timing Procedure

Step 1: Positioning the Needle Bar
• Set the needle bar at its lowest position
• Adjust so the eye of the needle is about 3mm above the throat plate when rising

Step 2: Aligning the Hook Point
• Rotate the handwheel to bring the hook point in line with the center of the needle
• The hook point should be level with the top of the needle eye

Step 3: Setting the Hook Timing
• The hook point should pass the needle when the needle bar has risen 2.5mm from its lowest point
• Use timing gauges for precise measurement

Step 4: Adjusting Feed Dog Timing
• Ensure feed dogs are below the throat plate when the needle enters the fabric
• Adjust so feed dogs reach their highest point just as the needle exits the fabric

Troubleshooting Tip: Always rotate the handwheel manually through several complete cycles after adjusting timing to ensure smooth operation and no unexpected obstructions.

4. Specialized Timing Considerations

a) For Zigzag Machines:
- Time the hook when the needle is at its rightmost position
- Ensure proper clearance between needle and hook at all needle positions

b) For Overlock Machines:
- Time multiple loopers in relation to each other and the needles
- Adjust needle guards to prevent needle deflection

c) For Chain Stitch Machines:
- Time the looper to pass behind the needle at the correct moment
- Adjust looper height for proper thread pickup

5. Common Timing Issues and Solutions

a) Skipped Stitches:
- Cause: Hook passing too early or too late
- Solution: Adjust hook timing precisely

b) Thread Breakage:
- Cause: Hook too close to needle or incorrect needle bar height
- Solution: Adjust clearance and check needle bar height

c) Fabric Damage:
- Cause: Incorrect feed dog timing or height
- Solution: Adjust feed dog timing and height

Troubleshooting Tip: When addressing timing issues, make small, incremental adjustments. Test after each adjustment to avoid over-correction.

6. Adjusting Heavy-Duty Machines

Beyond timing, heavy-duty machines require various adjustments for optimal performance:

a) Tension Adjustment:
 • Upper Tension: Adjust for different thread types and fabric weights
 • Bobbin Tension: Set slightly looser than upper tension
 • Check Springs: Adjust for proper thread control

b) Presser Foot Pressure:
 • Increase for heavy fabrics
 • Decrease for light or delicate materials

c) Stitch Length Adjustment:
 • Calibrate the stitch length regulator
 • Ensure consistency across the full range of lengths

d) Feed Dog Height:
 • Adjust so dogs are just visible above the throat plate at their highest point
 • Higher for thick fabrics, lower for thin materials

Troubleshooting Tip: Keep a record of optimal settings for different materials and projects. This can save significant time when switching between different types of work.

7. Specialized Adjustments for Industrial Machines

a) Needle Bar Height:
 • Critical for proper hook timing
 • Usually set so the needle eye is just below the hook point at its lowest position

b) Hook Gib Adjustment:
- Controls the gap between the hook and bobbin case
- Too tight causes drag, too loose affects stitch quality

c) Puller Feed Adjustment (if applicable):
- Synchronize with the main feed mechanism
- Adjust pressure for different fabric types

d) Walking Foot Timing (for walking foot machines):
- Coordinate movement with needle and feed dog action
- Adjust for different fabric thicknesses

8. Tools for Timing and Adjustment

- Timing Gauges: For precise measurement of component positions
- Screwdrivers: Various sizes for different adjustments
- Tension Gauge: To measure and set accurate thread tensions
- Stroboscope: For checking timing at high speeds

9. Advanced Timing Techniques

- Using Slow Motion Video: Record machine operation to analyze movement
- Marking Components: Use timing marks for visual reference
- Test Sewing: Perform test runs on various fabrics after each adjustment

Troubleshooting Tip: When working on unfamiliar machines, take photos or videos before disassembly. This can be invaluable when reassembling and provides a reference point for timing.

10. Safety Considerations

- Always disconnect power before making internal adjustments
- Be cautious of sharp needles and hooks
- Use proper lighting to avoid eye strain during precise adjustments

11. Maintaining Timing and Adjustments

- Regularly check and retighten all adjustment screws
- Perform timing checks after replacing major components
- Consider creating a maintenance schedule for each machine

12. Troubleshooting Complex Timing Issues

- Isolate problems by testing with different threads and fabrics
- Use process of elimination to identify the root cause
- Consult manufacturer manuals for model-specific troubleshooting guides

Timing and adjusting heavy-duty sewing machines requires patience, precision, and a deep understanding of machine mechanics. By mastering these skills, you can ensure that industrial machines operate at peak efficiency, producing high-quality results consistently.

Remember, every adjustment can have a ripple effect on machine performance. Always approach timing and adjustment methodically, making small changes and testing thoroughly after each modification. With practice and attention to detail, you'll develop the expertise to keep even the most complex industrial sewing machines running smoothly.

In our next section, we'll explore how to solve industrial-specific issues that go beyond timing and basic adjustments. Are you ready to tackle the unique challenges presented by specialized industrial sewing equipment?

Solving Industrial-Specific Issues

Industrial sewing machines, while robust and designed for high-volume production, can encounter specific issues due to their specialized nature and intense usage. Understanding and addressing these problems is crucial for maintaining productivity and quality in industrial settings.

1. High-Speed Operation Issues

Industrial machines often run at speeds of 5000 stitches per minute or more, leading to unique problems:

a) Excessive Heat Generation:
 • Symptom: Machine becomes hot to touch, burning smell
 • Solution:
  - Check and upgrade cooling systems if necessary
  - Ensure proper lubrication
  - Consider installing additional cooling fans

b) Vibration at High Speeds:
 • Symptom: Machine shakes or produces excessive noise
 • Solution:
  - Check and tighten all mounting bolts
  - Inspect for worn bearings or bushings
  - Ensure proper balance of rotating parts

Troubleshooting Tip: Regularly use a infrared thermometer to monitor machine temperature during operation. This can help identify heat-related issues before they cause damage.

2. Heavy Material Handling Problems

Industrial machines often work with thick, heavy fabrics, leading to specific issues:

a) Needle Breakage:
  • Symptom: Frequent needle snapping, especially with thick materials
  • Solution:
  - Use appropriate needle size and type for the material
  - Adjust needle bar height and timing
  - Check for any burrs on the throat plate or feed dogs

b) Inadequate Penetration:
  • Symptom: Skipped stitches or incomplete penetration on thick materials
  • Solution:
  - Increase presser foot pressure
  - Use a walking foot or compound feed system
  - Adjust thread tension for heavier materials

Troubleshooting Tip: Keep a log of successful settings for different material types. This can save significant setup time when switching between jobs.

3. Specialized Stitch Formation Issues

Many industrial machines are designed for specific stitch types, each with their own potential problems:

a) Chainstitch Unraveling:
  • Symptom: Stitches come undone easily
  • Solution:
  - Adjust looper timing and position
  - Check thread tension, especially on the looper thread

- Ensure proper thread path and lubrication

b) Overlock Stitch Balancing:
• Symptom: Uneven or loose overlock stitches
• Solution:
- Balance tensions between needle and looper threads
- Adjust differential feed settings
- Check knife position and sharpness

Troubleshooting Tip: Use contrasting thread colors in different components (e.g., needle, upper looper, lower looper) to easily identify which part of the stitch is problematic.

4. Automated Feature Malfunctions

Modern industrial machines often include automated features that can malfunction:

a) Automatic Thread Trimmers:
• Symptom: Failure to cut thread or incomplete cutting
• Solution:
- Check and sharpen trimmer blades
- Adjust trimmer timing and position
- Clean any lint or debris from the trimmer mechanism

b) Programmable Stitch Patterns:
• Symptom: Incorrect stitch patterns or machine not following programming
• Solution:
- Check and update machine software if applicable
- Verify and reinstall stitch pattern data
- Inspect stepper motors and solenoids for proper operation

Troubleshooting Tip: Keep a backup of all programmed stitch patterns and machine settings. This can be crucial for quick recovery if electronic components fail.

5. Industrial-Specific Lubrication Problems

Proper lubrication is critical in high-speed industrial machines:

a) Oil Leakage:
  • Symptom: Oil spots on fabric, excessive oil accumulation
  • Solution:
  - Check and replace worn seals or gaskets
  - Adjust oil flow regulators if available
  - Ensure proper oil viscosity for the machine

b) Insufficient Lubrication:
  • Symptom: Increased noise, heat, or wear on moving parts

  • Solution:
  - Clean and unclog oil passages
    Check oil pump operation
  - Adjust oil levels and distribution settings

Troubleshooting Tip: Implement a regular maintenance schedule that includes checking and changing oil. Use only manufacturer-recommended lubricants to avoid compatibility issues.

6. Power Transmission Problems

Industrial machines often use powerful motors and complex power transmission systems:

a) Belt Issues:
• Symptom: Slipping, fraying, or breaking belts
• Solution:
- Check and adjust belt tension
- Replace worn or damaged belts
- Ensure proper pulley alignment

b) Clutch and Brake Malfunctions:
• Symptom: Delayed stopping or starting, machine doesn't stop when pedal is released
• Solution:
- Adjust clutch and brake mechanisms
- Replace worn clutch or brake pads
- Check and adjust pedal linkage

Troubleshooting Tip: Listen for unusual sounds during operation. A well-functioning machine should have a consistent sound; any changes can indicate developing issues.

7. Feed Mechanism Complications

Industrial machines often have specialized feed systems:

a) Puller Feed Synchronization:
• Symptom: Fabric puckering or stretching
• Solution:
- Adjust puller feed speed to match main feed dogs
- Check for wear on puller feed wheels
- Synchronize puller feed timing with needle movement

b) Differential Feed Issues:
- Symptom: Wavy seams or fabric gathering
- Solution:
- Adjust differential feed ratio
- Check for worn or damaged feed dogs
- Ensure proper presser foot pressure

Troubleshooting Tip: Test adjustments on scrap material that matches your production fabric. Small changes in feed settings can have significant effects on different materials.

8. Electrical and Electronic Issues

Modern industrial machines often incorporate sophisticated electronics:

a) Control Panel Malfunctions:
- Symptom: Unresponsive buttons, error messages
- Solution:
- Check and reseat internal connections
- Update software or firmware if available
- Replace faulty control panels

b) Stepper Motor Problems:
- Symptom: Erratic needle or feed movement
- Solution:
- Check motor wiring and connections
- Verify motor driver functionality
- Replace faulty stepper motors

Troubleshooting Tip: Keep a multimeter handy for quick electrical checks. Understanding basic electronics can be invaluable for troubleshooting modern industrial machines.

9. Environmental Factors

Industrial settings can expose machines to challenging conditions:

a) Dust and Lint Accumulation:
- Symptom: Increased wear, stitch quality issues
- Solution:
- Implement regular cleaning schedules
- Use compressed air to blow out hard-to-reach areas
- Consider installing additional filtration systems

b) Humidity and Temperature Fluctuations:
- Symptom: Rusting components, electronic malfunctions
- Solution:
- Control workshop environment where possible
- Use dehumidifiers in humid environments
- Ensure proper machine warm-up in cold conditions

Troubleshooting Tip: Create a checklist for daily, weekly, and monthly maintenance tasks. Consistent care can prevent many environment-related issues.

Solving industrial-specific sewing machine issues requires a combination of mechanical knowledge, electrical understanding, and practical experience. By familiarizing yourself with these common problems and their solutions, you'll be better equipped to keep industrial sewing operations running smoothly and efficiently.

Remember, prevention is often the best cure. Regular maintenance, careful monitoring, and prompt attention to small issues can prevent many major problems in industrial sewing machines. Always prioritize safety, and don't hesitate to consult manufacturer resources or professional technicians for complex issues beyond your expertise.

In our next section, we'll explore the world of serger and overlocker maintenance, diving into the unique challenges presented by these specialized machines. Are you ready to tackle the intricacies of multi-thread sewing systems?

Chapter 6
Serger and Overlocker Maintenance
Mastering Serger Threading

Sergers and overlockers are essential machines in many sewing operations, known for their ability to create professional-looking finished edges and seams. However, their multi-thread systems can be intimidating. Mastering the threading process is crucial for proper operation and maintenance.

1. Understanding Serger Anatomy

Before threading, familiarize yourself with key components:
- Loopers (upper and lower)
- Needles (typically 2-4)
- Thread guides and tension dials
- Thread tree or stand
- Knife or blade system

Troubleshooting Tip: Create a color-coded diagram of your serger's threading path. This visual aid can be invaluable when learning or refreshing your memory.

2. General Threading Principles

a) Thread Quality:
- Use high-quality, smooth thread designed for sergers
- Ensure all threads are of the same fiber content for balanced tension

b) Thread Path:
- Always thread from back to front
- Follow manufacturer's recommended threading order

c) Tension Settings:
- Start with all tensions at neutral settings
- Adjust individually after threading

Troubleshooting Tip: If you're experiencing consistent issues, try rethreading the entire machine. Many problems stem from incorrect threading.

3. Step-by-Step Threading Process

Step 1: Prepare the Machine
- Turn off the machine for safety
- Raise the presser foot to release tension discs
- Extend the thread tree fully

Step 2: Lower Looper Threading
- Often the most challenging, thread this first
- Guide thread through designated points, ending at the looper eye
- Use tweezers for hard-to-reach areas

Step 3: Upper Looper Threading
- Follow a similar path to the lower looper
- Ensure thread is seated correctly in all guides

Step 4: Right Needle Threading
- Thread from back to front
- Ensure thread is between tension discs
- Use needle threader if available

Step 5: Left Needle Threading (if applicable)
- Follow the same principle as the right needle
- Double-check that threads aren't crossed

Troubleshooting Tip: If threads keep breaking during threading, check for rough spots on thread guides or the needle eye. Use a fine emery board to gently smooth any burrs.

4. Specialized Threading Techniques

a) Air Threading Systems:
 • Ensure air compressor is functioning properly
 • Keep air holes clean and free from lint
 • Follow manufacturer's specific instructions for air threading

b) Chain Stitch Loopers:
 • Often require a different threading path
 • Pay extra attention to looper positioning

c) Coverstitch Machines:
 • May have additional thread guides
 • Often require threading loopers from front to back

5. Common Threading Issues and Solutions

a) Thread Breaks During Threading:
 • Cause: Rough spots on guides or incorrect path
 • Solution: Check and smooth thread path, ensure correct routing

b) Difficulty Reaching Looper Eye:
 • Cause: Limited visibility or access
 • Solution: Use threading wire or specialized tools provided with the machine

c) Threads Tangling:
  • Cause: Incorrect threading order or crossed threads
  • Solution: Follow recommended threading order, ensure threads are separate

Troubleshooting Tip: If you're consistently struggling with a particular threading point, use a magnifying glass to inspect the area. There might be a small burr or damage causing the issue.

6. Maintenance Tips for Easy Threading

- Clean the machine thoroughly after each use
- Oil threading points as recommended by the manufacturer
- Replace needles regularly to prevent thread shredding
- Keep threading tools organized and easily accessible

7. Advanced Threading Techniques

a) Speed Threading:
  • Tie new thread to old thread and pull through
  • Useful for quick color changes
  • Ensure knots are small enough to pass through all guides

b) Threading for Specialty Threads:
  • Use thread nets for slippery threads
  • Adjust tension for thicker or metallic threads
  • Consider using a separate stand for large cones

8. Troubleshooting Thread Tension After Threading

- Test stitch on scrap fabric after threading
- Adjust tensions one at a time, starting with loopers
- Refer to your manual for ideal stitch appearance

Troubleshooting Tip: Create a "tension diary" noting ideal settings for different fabrics and threads. This can save significant time in future projects.

9. Safety Considerations

- Always turn off the machine before threading
- Be cautious around the knife area
- Use proper lighting to reduce eye strain

10. Teaching and Learning Strategies

- Practice threading with contrasting thread colors
- Use numbered stickers to mark threading points
- Consider filming yourself threading for future reference

11. Maintaining Threading Efficiency

- Regularly check and replace worn thread guides
- Keep a clean, organized workspace for easy access to tools
- Practice threading regularly to maintain muscle memory

Mastering serger threading is a skill that comes with practice and patience. Remember, even experienced sewists sometimes need to refer back to their machine's manual. Don't get discouraged if it takes a few attempts to get it right – each time you thread your serger, you're building valuable skills and knowledge.

By understanding the principles behind serger threading and practicing these techniques, you'll be well-equipped to handle even the most complex serging tasks. Remember, a well-threaded serger is the foundation for beautiful, professional-looking seams and edges.

In our next section, we'll explore the intricacies of blade replacement and adjustment in sergers and overlockers. Are you ready to ensure your machine cuts as perfectly as it sews?

Blade Replacement and Adjustment

The cutting blade, or knife, is a critical component of sergers and overlockers. It trims the fabric edge while sewing, creating a clean, professional finish. Proper maintenance and adjustment of this system are essential for optimal performance.

1. Understanding the Blade System

Sergers typically have two blades:
- Upper blade (movable)
- Lower blade (fixed or stationary)

These work together like scissors to trim fabric edges.

Troubleshooting Tip: If you notice frayed fabric edges or skipped stitches, it's often a sign that your blades need attention.

2. Signs That Blade Replacement is Needed

- Fabric edges appear frayed or chewed rather than cleanly cut
- Increased fabric build-up around the blade area
- Visible nicks or dull spots on the blade edge
- Machine struggling to cut through fabric

3. Blade Replacement Process

Step 1: Safety First
- Unplug the machine
- Wear protective gloves to avoid cuts

Step 2: Access the Blade Area
- Open the looper cover
- Remove the stitch finger or seam guide if necessary

Step 3: Remove the Old Blade
- For the upper blade:
- Loosen the securing screw
- Carefully lift out the old blade
- For the lower blade (if replaceable):
- This often requires removing multiple screws
- Note exact positioning before removal

Step 4: Install the New Blade
- Insert the new blade, ensuring it's seated correctly
- Tighten screws, but don't fully tighten yet

Step 5: Adjust Blade Position
- Align upper blade with lower blade
- Ensure blades meet at the correct point (usually just behind the needle plate)

Step 6. Test and Fine-tune
- Manually rotate the handwheel to check blade interaction
- Make fine adjustments as needed
- Fully tighten all screws once position is correct

Troubleshooting Tip: If you're unsure about blade positioning, consult your machine's manual or mark the old blade's position before removal as a reference.

4. Blade Adjustment Techniques

a) Adjusting Cutting Width:
- Locate the adjustment screw (usually on the upper blade)
- Turn clockwise to move blade closer to needle plate (narrower cut)
- Turn counterclockwise for a wider cut
- Aim for about 1/8 inch (3.5mm) of fabric trimming

b) Aligning Upper and Lower Blades:
- Upper blade should slightly overlap lower blade
- Adjust so blades make contact along their entire length
- Check alignment by slowly turning the handwheel

c) Setting Blade Pressure:
- Some machines allow adjustment of how firmly blades press together
- Increase pressure for thicker fabrics
- Decrease for delicate materials to prevent crushing

Troubleshooting Tip: After any blade adjustment, always test on scrap fabric similar to your project material. This ensures the adjustments are suitable for your specific needs.

5. Maintaining Blade Sharpness

- Clean blades regularly with a small brush to remove lint and debris
- Use a sharpening stone designed for serger blades (follow manufacturer instructions)
- Consider professional sharpening for best results

6. Troubleshooting Common Blade Issues

a) Blades Not Cutting Cleanly:
• Cause: Dull blades or incorrect alignment
• Solution: Sharpen or replace blades, check and adjust alignment

b) Fabric Being Pushed Rather Than Cut:
• Cause: Insufficient blade pressure or very dull blades
• Solution: Adjust blade pressure, replace blades if necessary

c) Uneven Cutting:
• Cause: Blade misalignment or damage
• Solution: Carefully realign blades, check for nicks or damage

d) Loud Clicking or Grinding Noise:
• Cause: Blades touching too aggressively or foreign object present
• Solution: Readjust blade alignment, check for and remove any debris

7. Specialized Blade Considerations

a) Curved Blades:
• Some sergers use curved blades for improved cutting
• Require extra care in alignment and replacement

b) Non-Cutting Serging:
• Some projects require serging without trimming
• Learn how to disengage the upper blade on your machine

8. Safety and Maintenance Tips

- Always store spare blades in a safe, dry place
- Oil the blade mechanism as recommended by the manufacturer
- Never force the blades if you feel resistance during adjustment

9. Extending Blade Life

- Use the appropriate needle size and thread weight for your fabric
- Avoid serging over pins or other hard objects
- Consider disengaging blades when not needed (e.g., for rolled hems)

Troubleshooting Tip: If you frequently work with a variety of fabric weights, consider keeping multiple sets of blades, each sharpened for specific material types.

10. When to Seek Professional Help

- If you're uncomfortable performing blade replacement yourself
- When dealing with specialized or industrial sergers
- If you've made multiple adjustments without resolving the issue

11. Blade Care for Different Fabric Types

- Use fresh, sharp blades for delicate fabrics to prevent snagging
- Slightly dull blades can sometimes work better on very thick materials

- Consider dedicated blades for problematic fabrics like vinyl or leather

Proper blade maintenance and adjustment are crucial for the optimal performance of your serger or overlocker. By mastering these skills, you'll ensure clean, precise cuts and professional-looking finished edges on all your projects.

Remember, patience and attention to detail are key when working with serger blades. Always prioritize safety, and don't hesitate to consult your machine's manual or a professional if you're unsure about any aspect of blade replacement or adjustment.

In our next section, we'll explore the intricacies of tension troubleshooting for perfect stitches. Are you ready to achieve flawless serger seams every time?

Tension Troubleshooting for Perfect Stitches

Achieving perfect tension on a serger or overlocker is often considered an art form. It's crucial for creating strong, balanced stitches that look great on both sides of the fabric. Let's explore how to troubleshoot and perfect your serger's tension.

1. Understanding Serger Tension

Sergers typically have 3-5 tension dials controlling:
- Left needle thread
- Right needle thread
- Upper looper thread
- Lower looper thread
- Chain stitch thread (if applicable)

Each thread must work in harmony for a perfect stitch.

Troubleshooting Tip: Always start with all tensions at neutral (usually marked on the machine) before making adjustments.

2. Recognizing Perfect Tension

A well-balanced serger stitch should:
- Lay flat without puckering
- Have evenly spaced needle threads on the top
- Show looper threads meeting exactly at the fabric edge
- Look similar on both sides of the fabric

3. Common Tension Issues and Solutions

a) Loose Upper Looper Thread:
- Symptom: Loops on the underside of the fabric
- Solution: Increase upper looper tension

b) Loose Lower Looper Thread:
- Symptom: Loops on the top side of the fabric
- Solution: Increase lower looper tension

c) Needle Threads Too Tight:
- Symptom: Fabric puckers or breaks needle thread
- Solution: Decrease needle thread tension

d) Needle Threads Too Loose:
- Symptom: Large loops on the edge of the fabric
- Solution: Increase needle thread tension

Troubleshooting Tip: Make small, incremental adjustments. It's easier to fine-tune than to start over after a drastic change.

4. Step-by-Step Tension Adjustment Process

Step 1: Prepare a Test Swatch
- Use fabric similar to your project
- Mark which side is the right side

Step 2: Start with Looper Threads
- Adjust lower looper first, then upper looper
- Aim for looper threads to meet at the fabric edge

Step 3: Adjust Needle Threads
- Balance left and right needle tensions
- Ensure stitches are even and flat

Step 4: Fine-Tune All Tensions
- Make minor adjustments to all threads for perfect balance
- Test after each adjustment

5. Factors Affecting Tension

- Thread type and quality
- Fabric weight and texture
- Stitch length and width settings
- Differential feed settings

Troubleshooting Tip: Keep a "tension diary" noting ideal settings for different fabric and thread combinations. This can save time on future projects.

6. Advanced Tension Troubleshooting

a) Stitch Fingers:
- Adjust or replace if stitches are inconsistent
- Ensure they're not bent or damaged

b) Thread Path:
- Verify threads are correctly seated in all guides and tension discs
- Check for any obstructions or rough spots

c) Needle Issues:
- Replace bent or dull needles
- Ensure needles are inserted correctly and fully seated

7. Tension for Specialty Techniques

a) Rolled Hems:
- Typically require higher upper looper tension
- May need to disengage one needle

b) Flatlock Stitching:
• Requires intentional imbalance in tensions
• Experiment with dramatically different upper and lower looper tensions

c) Stretch Fabrics:
• Often need looser tensions overall
• Adjust differential feed in conjunction with tension

Troubleshooting Tip: When working with specialty techniques, always test on scraps until you achieve the desired effect before moving to your actual project.

8. Maintenance for Consistent Tension

• Clean tension discs regularly with canned air or a soft brush
• Oil machine as recommended by the manufacturer
• Replace threads if they appear worn or inconsistent

9. Troubleshooting Persistent Tension Problems

a) Thread Breaking:
• Cause: Tension too tight or rough spot in thread path
• Solution: Decrease tension, check and smooth thread path

b) Skipped Stitches:
• Cause: Incorrect threading or needle issues
• Solution: Rethread machine, replace needle

c) Wavy Edges:
• Cause: Differential feed setting incorrect
• Solution: Adjust differential feed in conjunction with tension

10. Using Different Thread Types

- Wooly Nylon: Often used in loopers, may require looser tension
- Decorative Threads: May need tension adjustments and use of thread nets
- All-Purpose vs. Serger Thread: Serger-specific thread often performs better

11. Tension and Stitch Length Relationship

- Shorter stitch lengths may require slightly tighter tensions
- Longer stitch lengths often need looser tensions
- Always re-check tension after changing stitch length

Troubleshooting Tip: If you're having trouble achieving balance, try slightly changing your stitch length. Sometimes a small adjustment here can make tension balancing easier.

12. Visual Aids for Tension Troubleshooting

- Create a reference card with diagrams of common tension issues
- Use different colored threads in each component to easily identify which tension needs adjusting

13. When to Seek Professional Help

- If tensions won't stay consistent after adjustment
- When you suspect internal tension assembly issues
- If you're working with an unfamiliar or complex serger model

Remember, achieving perfect tension on a serger is a skill that develops with practice. Don't get discouraged if it takes some time to master. Each fabric and thread combination may require slight adjustments, but with patience and systematic troubleshooting, you'll be creating flawless serged seams and edges in no time.

Always keep your manual handy, as it often contains model-specific tension troubleshooting tips. And don't be afraid to experiment on scrap fabric – it's the best way to understand how your particular machine responds to different tension adjustments.

In our next section, we'll explore advanced repair techniques for sewing machines, taking your maintenance and repair skills to the next level. Are you ready to become a true sewing machine repair expert?

Chapter 7
Advanced Repair Techniques

Motor Repair and Replacement

The motor is the heart of any sewing machine, providing the power to drive all mechanical functions. Understanding how to repair and replace motors is a valuable skill for any serious sewing machine technician.

1. Understanding Sewing Machine Motors

Types of motors commonly found in sewing machines:
- Universal Motors: Most common in domestic machines
- Servo Motors: Often found in modern computerized machines
- DC Motors: Used in some portable or specialized machines

Components of a typical sewing machine motor:
- Armature (rotor)
- Field coils (stator)
- Brushes
- Commutator
- End caps
- Cooling fan

Troubleshooting Tip: Before diving into motor repair, always check for simpler issues like loose connections or a faulty foot pedal.

2. Common Motor Problems and Diagnostics

a) Motor Doesn't Run:
- Check power supply and connections

- Test the foot pedal with a multimeter
- Inspect brushes for wear

b) Motor Runs Slowly or Weakly:
- Check for worn brushes
- Inspect the commutator for wear or damage
- Test for weak field magnets

c) Motor Overheats:
- Check for obstructions in cooling vents
- Inspect bearings for wear
- Test for short circuits in windings

d) Excessive Noise:
- Look for loose or worn bearings
- Check for a bent shaft
- Inspect the cooling fan for damage

Troubleshooting Tip: Always start with the simplest possible cause. Many motor issues can be resolved without complete disassembly.

3. Basic Motor Maintenance

- Regular cleaning to remove dust and lint
- Proper lubrication of bearings (if applicable)
- Periodic brush inspection and replacement
- Checking and tightening of mounting screws

4. Advanced Motor Repair Techniques

a) Brush Replacement:
- Remove the motor from the machine
- Locate and remove brush caps

- Extract old brushes and springs
- Insert new brushes, ensuring proper alignment
- Reinstall brush caps

b) Commutator Repair:
- Remove the armature
- Use fine sandpaper to clean the commutator surface
- For deep grooves, use a commutator stone or have it professionally turned

c) Rewinding Field Coils:
- This is a complex process, often best left to professionals
- If attempting:
 - Carefully document the original winding pattern
 - Use the correct gauge of wire
 - Ensure proper insulation between windings

d) Bearing Replacement:
- Remove end caps
- Extract old bearings using a puller
- Press in new bearings, ensuring proper alignment
- Reinstall end caps

Troubleshooting Tip: When reassembling a motor, mark the position of end caps relative to the motor body. Incorrect alignment can cause issues with brush contact and overall performance.

5. Motor Replacement Process

Step 1: Identify the Correct Replacement Motor
- Match specifications (voltage, RPM, mounting style)
- Consider upgrading to a more efficient model if available

Step 2: Remove the Old Motor
- Disconnect wiring, noting connection points
- Remove mounting screws or bolts
- Carefully extract the motor

Step 3: Prepare the New Motor
- If necessary, transfer pulleys or gears from the old motor
- Ensure proper alignment of drive components

Step 4: Install the New Motor
- Mount the motor securely
- Reconnect wiring, following your earlier notes
- Adjust belt tension if applicable

Step 5: Test and Fine-tune
- Run the machine at various speeds
- Check for unusual noises or vibrations
- Adjust mounting if necessary for smooth operation

Troubleshooting Tip: Take photos of the original motor installation before removal. This visual reference can be invaluable during replacement.

6. Specialized Motor Considerations

a) Servo Motors:
- Often integrated with control boards
- May require programming or calibration after replacement
- Handle with care due to sensitive internal components

b) Variable Speed Motors:
- Check speed control module if motor runs at only one speed
- Ensure compatibility of replacement motor with existing speed control

c) Computerized Machine Motors:
- May have additional sensors or encoders
- Often require exact replacements for proper function

7. Safety Considerations

- Always unplug the machine before working on the motor
- Discharge any capacitors before handling
- Use insulated tools when working with electrical components
- Wear safety glasses to protect from carbon dust when working with brushes

8. Troubleshooting Complex Motor Issues

a) Intermittent Operation:
- Check for loose internal connections
- Inspect the centrifugal switch if present
- Look for hairline cracks in wiring insulation

b) Excessive Sparking at Brushes:
- Indicative of commutator issues or misaligned brushes
- May require professional commutator resurfacing

c) Motor Runs But Machine Doesn't:
- Check clutch mechanism
- Inspect for stripped gears in the power transmission system

9. When to Replace vs. Repair

Consider replacement when:
- Cost of repairs approaches 50% of a new motor's price
- Original motor is obsolete or parts are unavailable

- Upgrading to a more efficient motor would provide significant benefits

10. Tools for Motor Repair and Replacement

- Multimeter for electrical testing
- Screwdrivers and nut drivers
- Bearing puller
- Soldering iron for minor wiring repairs
- Commutator stone

11. Preventive Measures for Motor Longevity

- Avoid overloading the machine
- Ensure proper ventilation around the motor
- Use the machine regularly to prevent bearing issues from disuse
- Keep the machine and motor clean and dust-free

Troubleshooting Tip: If you frequently work in dusty environments, consider creating a simple filter for the motor's cooling vents using fine mesh or foam.

Motor repair and replacement can be complex, but with patience and the right approach, many issues can be resolved. Always prioritize safety and don't hesitate to seek professional help for repairs beyond your comfort level or expertise.

Remember, a well-maintained motor is key to the smooth operation of your sewing machine. Regular checks and prompt attention to minor issues can prevent major breakdowns and extend the life of your machine significantly.

In our next section, we'll explore the intricacies of circuit board diagnostics and fixes, delving deeper into the electronic heart of modern sewing machines. Are you ready to blend your mechanical skills with some electronic troubleshooting?

Circuit Board Diagnostics and Fixes

As sewing machines become more computerized, understanding circuit board issues becomes crucial for advanced repairs. This section will cover how to diagnose and potentially fix common circuit board problems.

1. Understanding Sewing Machine Circuit Boards

Components typically found on sewing machine circuit boards:
- Microprocessors
- Capacitors
- Resistors
- Transistors
- Diodes
- Integrated Circuits (ICs)
- Connectors and Sockets

Troubleshooting Tip: Before diving into circuit board repairs, always check for simpler issues like loose connections or software glitches that might mimic hardware problems.

2. Common Circuit Board Issues

a) Power Supply Problems:
- Symptoms: Machine doesn't turn on, intermittent operation
- Possible causes: Faulty capacitors, damaged voltage regulators

b) Sensor Malfunctions:
- Symptoms: Erratic machine behavior, unresponsive features
- Possible causes: Damaged sensor connections, faulty sensor ICs

c) Motor Control Issues:
- Symptoms: Motor doesn't run or runs erratically
- Possible causes: Damaged motor driver chips, blown fuses

d) Display Problems:
- Symptoms: Blank or garbled LCD display
- Possible causes: Faulty display connector, damaged display driver

3. Basic Diagnostic Tools and Techniques

Essential tools:
- Multimeter
- Magnifying glass or loupe
- Soldering iron and desoldering pump
- Anti-static wrist strap

Basic diagnostic steps:
1. Visual inspection for obvious damage
2. Checking for loose connections
3. Testing voltages at key points
4. Continuity testing of suspicious components

Troubleshooting Tip: Always work in a well-lit, static-free environment. Even a small static discharge can damage sensitive components.

4. Advanced Diagnostic Techniques

a) Signal Tracing:
- Use an oscilloscope to follow signal paths
- Compare readings to expected waveforms (if available in service manual)

b) Thermal Imaging:
- Use a thermal camera to identify hot spots on the board
- Overheating components often indicate failure

c) Logic Analyzer:
- For complex digital circuits
- Helps diagnose timing and data transfer issues

5. Common Circuit Board Fixes

a) Resoldering Loose Connections:
1. Identify the loose joint
2. Clean the area with isopropyl alcohol
3. Apply flux and resolder using appropriate temperature

b) Capacitor Replacement:
1. Identify the faulty capacitor (often bulging or leaking)
2. Desolder and remove the old capacitor
3. Replace with an identical or equivalent capacitor
4. Resolder, ensuring correct polarity

c) Fuse Replacement:
1. Locate the blown fuse (often visibly damaged)
2. Remove and test with a multimeter to confirm
3. Replace with a fuse of the same rating

d) Cleaning Corrosion:
1. Gently scrub affected areas with a soft brush and isopropyl alcohol
2. For severe corrosion, use a specialized electronics cleaner
3. Dry thoroughly before powering on

Troubleshooting Tip: When replacing components, always double-check values and polarity. A simple mistake can cause further damage.

6. Dealing with Integrated Circuits (ICs)

- Diagnosing faulty ICs often requires specialized equipment
- In many cases, it's more practical to replace the entire board
- If replacement is necessary:
  1. Ensure exact match of IC type and specifications
  2. Use proper desoldering and resoldering techniques
  3. Consider using a socket for easy future replacement

7. Software and Firmware Issues

- Some apparent hardware issues may be software-related
- Check for available firmware updates
- Consider performing a factory reset if possible

Troubleshooting Tip: Keep a log of any error codes displayed. These can be crucial in diagnosing both hardware and software issues.

8. When to Replace vs. Repair

Consider board replacement when:
- Multiple components are damaged
- The cost of individual repairs exceeds replacement cost
- The board shows signs of extensive heat damage or corrosion

9. Preventive Measures

- Keep the machine in a clean, dust-free environment
- Use surge protectors to prevent power-related damage
- Avoid exposing the machine to extreme temperatures or humidity

10. Safety Considerations

- Always unplug the machine before working on circuit boards
- Use an anti-static wrist strap to prevent static discharge
- Be cautious of residual charges in capacitors

11. Advanced Troubleshooting Scenarios

a) Intermittent Issues:
- Often the most challenging to diagnose
- Look for hairline cracks in solder joints
- Check for loose connectors that may shift during operation

b) Multiple Fault Symptoms:
- Start by addressing one issue at a time
- Be aware that fixing one problem may reveal others

c) No Power to Specific Functions:
- Trace the power path to the affected components
- Check for blown fuses or damaged voltage regulators

Troubleshooting Tip: When dealing with complex issues, create a flowchart of the diagnostic process. This can help organize your thoughts and ensure you don't overlook any steps.

12. Resources for Circuit Board Repair

- Manufacturer service manuals (if available)
- Online forums and communities dedicated to sewing machine repair
- Electronics repair courses or workshops

Remember, circuit board repair requires patience, precision, and often specialized knowledge. While many issues can be resolved with careful troubleshooting and basic repairs, don't hesitate to seek professional help for complex problems or if you're unsure about any aspect of the repair process.

By understanding the basics of circuit board diagnostics and repairs, you'll be better equipped to tackle a wide range of issues in modern computerized sewing machines. However, always prioritize safety and be aware of your limitations when working with sensitive electronic components.

In our next section, we'll explore tension assembly overhaul, a critical aspect of maintaining stitch quality in all types of sewing machines. Are you ready to master the art of perfect thread tension?

Tension Assembly Overhaul

The tension assembly is vital for creating balanced, high-quality stitches. An overhaul of this system can resolve many common sewing issues and greatly improve your machine's performance.

1. Understanding the Tension Assembly

Key components typically include:
- Tension discs
- Tension spring
- Check spring
- Tension regulator
- Thread guide
- Pretension guide

Troubleshooting Tip: Before overhauling the tension assembly, always check for simpler issues like incorrect threading or a dirty bobbin area, which can mimic tension problems.

2. Signs That a Tension Assembly Overhaul is Needed

- Inconsistent stitch quality
- Frequent thread breakage
- Inability to adjust tension effectively
- Visible wear or damage to tension components
- Strange noises when thread passes through the assembly

3. Preparation for Overhaul

Tools needed:
- Small screwdrivers (flathead and Phillips)
- Tweezers
- Cleaning brushes

- Compressed air (optional)
- Lubricant suitable for your machine
- Replacement parts (if necessary)

Before starting:
- Take clear photos of the assembly for reference
- Note the current tension settings

Troubleshooting Tip: Create a small parts tray or use a magnetic mat to keep tiny components organized during disassembly.

4. Step-by-Step Tension Assembly Overhaul Process

Step 1: Disassembly
- Remove the face plate or access panel
- Carefully unscrew the tension dial
- Remove tension discs, noting their order and orientation
- Extract the check spring and tension spring

Step 2: Cleaning
- Use compressed air or a soft brush to remove all lint and debris
- Clean each component thoroughly with a lint-free cloth
- Pay special attention to the spaces between tension discs

Step 3: Inspection
- Check tension discs for wear, scoring, or rust
- Inspect springs for loss of tension or damage
- Examine the tension post for bending or wear

Step 4: Repair or Replacement
- Replace any worn or damaged parts
- If tension discs are scored, consider carefully polishing with fine emery cloth

Step 5: Lubrication
- Apply a tiny amount of suitable lubricant to the tension post
- Ensure no lubricant gets on the tension discs

Step 6: Reassembly
- Reinstall components in the reverse order of disassembly
- Ensure proper alignment of all parts
- Be careful not to overtighten screws

Troubleshooting Tip: If you're unsure about the correct assembly order, refer to your photos or the machine's manual. Incorrect assembly can lead to serious tension issues.

5. Common Issues and Solutions

a) Tension Discs Not Releasing:
- Cause: Dirt build-up or misalignment
- Solution: Clean thoroughly and check for proper assembly

b) Check Spring Malfunction:
- Cause: Loss of tension or improper positioning
- Solution: Adjust or replace the check spring

c) Inconsistent Tension:
- Cause: Worn tension discs or weak tension spring
- Solution: Replace affected components

d) Thread Shredding:
- Cause: Burrs on thread path or over-tightened tension
- Solution: Smooth any rough spots and adjust tension

6. Fine-Tuning the Tension Assembly

- Start with the tension at its neutral setting
- Test on a double layer of medium-weight fabric
- Make small adjustments, testing after each change
- Aim for perfectly balanced stitches where top and bottom threads meet in the middle of the fabric

Troubleshooting Tip: Create a tension test sample using different colored threads for top and bobbin. This makes it easier to identify which thread needs adjustment.

7. Specialized Tension Considerations

a) For Computerized Machines:
- Some have electronically controlled tension
- May require recalibration after overhaul (consult manual)

b) For Sergers/Overlockers:
- Multiple tension units working together
- Adjust in a specific order: lower looper, upper looper, then needles

c) For Embroidery Machines:
- May have separate tension for embroidery functions
- Consider thread type when adjusting (e.g., metallic threads often need looser tension)

8. Preventive Maintenance for Tension Assemblies

- Clean the tension area regularly
- Avoid pulling thread backwards through the tension discs
- Use high-quality thread to reduce lint build-up
- Periodically check and adjust tension, even if no issues are apparent

9. Troubleshooting Complex Tension Issues

a) Tension Changes During Sewing:
• Check for burrs on thread path
• Ensure pretension guide is functioning correctly

b) Unable to Achieve Proper Tension:
• Verify bobbin tension is correct
• Check for worn or damaged feed dogs affecting fabric feed

c) Tension Issues with Specific Fabrics:
• Adjust presser foot pressure
• Consider using different needle types

Troubleshooting Tip: Keep a "tension diary" noting ideal settings for different fabric and thread combinations. This can save time on future projects.

10. When to Seek Professional Help

Consider professional servicing if:
• You're uncomfortable disassembling the tension assembly
• Issues persist after a thorough overhaul
• Your machine is under warranty

11. Advanced Tension Techniques

• Understanding thread physics (twist direction, fiber content)
• Balancing tension for decorative stitches
• Adjusting for specialty threads (e.g., monofilament, elastic thread)

Remember, overhauling the tension assembly requires patience and attention to detail. It's a delicate process, but mastering it can significantly improve your sewing results. Always refer to your machine's manual for model-specific guidance, and don't hesitate to seek professional help for complex issues.

By understanding and maintaining your machine's tension assembly, you'll be able to achieve consistent, high-quality stitches across a wide range of fabrics and projects. Regular maintenance and timely overhauls will keep your sewing machine performing at its best for years to come.

In our next section, we'll explore preventive maintenance strategies to keep your sewing machine running smoothly and avoid common issues before they arise. Are you ready to become a proactive sewing machine maintainer?

Chapter 8
Preventive Maintenance and Care
Creating a Maintenance Schedule

Regular maintenance is key to extending the life of your sewing machine and ensuring consistent, high-quality performance. A well-planned maintenance schedule can save you time, money, and frustration in the long run.

1. Understanding the Importance of Preventive Maintenance

Benefits of regular maintenance:
- Extends machine lifespan
- Ensures consistent stitch quality
- Prevents unexpected breakdowns
- Saves money on major repairs
- Improves overall sewing experience

Troubleshooting Tip: Keep a log of any unusual sounds, vibrations, or performance issues. These notes can be invaluable for identifying developing problems early.

2. Basic Components of a Maintenance Schedule

Every maintenance schedule should include:
- Daily/Per-Use Tasks
- Weekly Tasks
- Monthly Tasks
- Quarterly Tasks
- Annual Tasks

3. Daily/Per-Use Maintenance Tasks

After each use:
- Remove lint and thread debris from bobbin area
- Wipe down the exterior of the machine
- Cover the machine to protect from dust

Before each use:
- Check needle condition and replace if necessary
- Ensure proper threading
- Test stitch quality on scrap fabric

Troubleshooting Tip: Develop a habit of listening to your machine as you sew. Familiarizing yourself with its normal sounds will help you quickly identify when something's amiss.

4. Weekly Maintenance Tasks

- Clean feed dogs thoroughly
- Oil the machine (if recommended by manufacturer)
- Check and adjust tension if needed
- Inspect power cord for any damage

5. Monthly Maintenance Tasks

- Deep clean all accessible parts
- Check belt tension (for machines with visible belts)
- Inspect and clean bobbin case
- Test all stitch patterns and functions

6. Quarterly Maintenance Tasks

- Check and tighten all visible screws
- Inspect and clean internal mechanisms (if comfortable doing so)
- Lubricate gears and moving parts (as per manual instructions)
- Check alignment of needle and feed dogs

7. Annual Maintenance Tasks

- Perform a comprehensive internal cleaning
- Replace belts if showing signs of wear
- Check and replace worn out parts (e.g., needle plate, presser feet)
- Consider professional servicing for thorough inspection and adjustment

Troubleshooting Tip: Schedule your annual maintenance during a typically slow sewing period to minimize disruption to your projects.

8. Creating Your Personalized Maintenance Schedule

Step 1: Consult Your Machine's Manual
- Note any specific maintenance requirements
- Check recommended frequency for oiling and cleaning

Step 2: Assess Your Usage
- Heavy users may need more frequent maintenance
- Consider the types of fabrics and threads you commonly use

Step 3: Create a Calendar
• Use a physical calendar or digital tool
• Mark regular maintenance tasks
• Set reminders for less frequent tasks

Step 4: Keep a Maintenance Log
• Record date and type of maintenance performed
• Note any parts replaced or issues encountered

9. Adapting Your Schedule for Different Machine Types

a) Computerized Machines:
• Include software/firmware updates in your schedule
• Pay extra attention to cleaning sensors and electronic components

b) Sergers/Overlockers:
• Schedule more frequent cleaning due to lint generation
• Include regular blade checks and sharpening/replacement

c) Embroidery Machines:
• Include cleaning and lubricating the embroidery unit
• Schedule regular calibration checks

10. Tools and Supplies for Maintenance

Keep a maintenance kit ready with:
• Lint brush and tweezers
• Sewing machine oil
• Screwdrivers (various sizes)
• Cleaning cloths
• Compressed air canister
• Spare needles and bobbins

Troubleshooting Tip: Store your maintenance supplies in a dedicated box or bag to keep everything organized and easily accessible.

11. Incorporating Troubleshooting into Your Maintenance Routine

• During each maintenance session, perform a quick troubleshooting check
• Test for common issues like skipped stitches or tension problems
• Address any minor issues before they become major problems

12. Educating Other Users

If multiple people use the machine:
• Create a simple checklist for basic per-use maintenance
• Encourage users to report any unusual behavior or performance issues
• Consider holding a brief maintenance training session

13. Adjusting Your Schedule Based on Performance

• Be prepared to increase maintenance frequency if issues arise
• Reduce frequency of certain tasks if they consistently reveal no problems
• Always prioritize tasks that directly impact stitch quality and machine function

14. Seasonal Considerations

• In humid seasons, pay extra attention to rust prevention
• During dry seasons, be more vigilant about static electricity issues
• If the machine is stored unused for long periods, perform a thorough check before use

Troubleshooting Tip: If you live in an area with extreme temperature or humidity changes, consider using a dehumidifier or humidity monitor in your sewing area.

15. Professional Servicing

• Include professional servicing in your long-term maintenance plan
• Consider annual or bi-annual professional check-ups
• Always seek professional help for issues beyond your expertise or comfort level

Remember, a well-maintained sewing machine is a reliable partner in your creative endeavors. By creating and following a comprehensive maintenance schedule, you'll ensure that your machine remains in top condition, ready to tackle any project you have in mind.

Consistency is key in preventive maintenance. Even small, regular care tasks can significantly extend the life of your machine and improve its performance. Don't view maintenance as a chore, but as an investment in your sewing future.

In our next section, we'll explore proper cleaning and lubrication techniques to keep your machine running smoothly. Are you ready to dive into the nitty-gritty of sewing machine care?

Proper Cleaning and Lubrication Techniques

Keeping your sewing machine clean and well-lubricated is crucial for its smooth operation and longevity. Let's explore the best practices for these important maintenance tasks.

1. Understanding the Importance of Cleaning and Lubrication

Benefits of regular cleaning and lubrication:
- Prevents build-up of lint and debris
- Reduces wear on moving parts
- Ensures smooth operation
- Maintains stitch quality
- Extends the machine's lifespan

Troubleshooting Tip: If you notice any change in your machine's sound or performance, check for lint build-up or lack of lubrication first before assuming more serious issues.

2. Essential Cleaning Tools

Gather these tools for effective cleaning:
- Lint brush or small, soft paintbrush
- Tweezers
- Compressed air can (optional)
- Soft, lint-free cloths
- Cotton swabs
- Isopropyl alcohol (for stubborn grime)

c) Vintage Machines:
- May require more frequent oiling
- Be cautious of old, hardened grease in gears

3. Step-by-Step Cleaning Process

Step 1: Prepare the Machine
- Unplug the machine for safety
- Remove the needle, presser foot, and throat plate
- Open all accessible covers

Step 2: Remove Loose Debris
- Use the lint brush to sweep away loose lint and threads
- Pay special attention to the bobbin area and feed dogs

Step 3: Deep Clean
- Use tweezers to remove stubborn lint or thread pieces
- Gently use compressed air for hard-to-reach areas (keep the can upright)
- Clean the tension discs with a thin piece of lint-free cloth

Step 4: Wipe Down Surfaces
- Use a slightly damp cloth to wipe all surfaces
- For sticky residues, use a cloth lightly dampened with isopropyl alcohol

Step 5: Reassemble and Final Check
- Replace all removed parts
- Wipe down the exterior of the machine

Troubleshooting Tip: If you encounter a jammed area during cleaning, resist the urge to force it. Consult your manual or a professional to avoid causing damage.

4. Frequency of Cleaning

- Light users: Clean after every 5-10 hours of use
- Heavy users: Clean after every project or weekly
- Always clean before switching to a different fabric type

5. Understanding Sewing Machine Lubrication

Types of lubricants:
- Sewing machine oil (for most parts)
- Grease (for gears, typically in the machine's base)

Never use:
- WD-40 or other household lubricants
- Cooking oils or automotive oils

6. Essential Lubrication Tools

- Sewing machine oil
- Grease (if recommended by manufacturer)
- Long-nozzled oiler or syringe for precise application
- Lint-free cloth for wiping excess

7. Step-by-Step Lubrication Process

Step 1: Consult the Manual
- Identify lubrication points specific to your machine
- Note any areas that should not be oiled

Step 2: Apply Oil
- Place a drop of oil at each designated point
- Use a long-nozzled oiler for precision

Step 3: Run the Machine
- Operate the machine for a few minutes to distribute the oil

Step 4: Wipe Excess
- Use a clean cloth to remove any excess oil

Troubleshooting Tip: If you accidentally over-oil, run a scrap piece of absorbent fabric through the machine to soak up excess oil.

8. Frequency of Lubrication

- Follow manufacturer recommendations
- Typically, lubricate after every 8-10 hours of use
- Some modern machines require less frequent oiling

9. Special Considerations for Different Machine Types

a) Computerized Machines:
- Be extra cautious around electronic components
- Avoid getting oil on circuit boards or sensors

b) Sergers/Overlockers:
- Clean more frequently due to higher lint production
- Pay special attention to the knife area

10. Troubleshooting Common Cleaning and Lubrication Issues

a) Machine Runs Roughly After Cleaning:
- Ensure all parts are correctly reassembled
- Check for lint caught in moving parts

b) Oil Spots on Fabric:
- You may have over-oiled
- Run scrap fabric through the machine until spots disappear

c) Persistent Squeaking:
- Double-check all lubrication points
- Consider professional servicing if the issue persists

11. Environmental Considerations

• Work in a well-ventilated area when using cleaning solvents
• Dispose of oily rags properly to prevent fire hazards
• Consider using eco-friendly cleaning products when possible

12. Maintaining Your Cleaning and Lubrication Tools

• Clean brushes after each use
• Store oil and lubricants in a cool, dry place
• Replace cloths regularly to avoid reintroducing dirt

Troubleshooting Tip: If your machine is due for lubrication but you're out of sewing machine oil, it's better to wait until you can get the proper oil rather than using a substitute.

13. Creating a Cleaning and Lubrication Log

• Record dates of cleaning and lubrication
• Note any issues encountered or parts replaced
• Use this log to identify patterns and optimize your maintenance schedule

14. When to Seek Professional Help

• If you encounter any electrical issues during cleaning
• When dealing with internal mechanisms you're not familiar with
• If problems persist despite regular cleaning and lubrication

Remember, proper cleaning and lubrication are fundamental to your sewing machine's health. These tasks may seem tedious, but they're invaluable in preventing more serious issues and ensuring your machine performs at its best.

Develop a routine that works for you and stick to it. Over time, these maintenance tasks will become second nature, and you'll reap the benefits of a well-maintained, smooth-running sewing machine for years to come.

In our next section, we'll explore storage and transportation tips to protect your sewing machine when it's not in use or on the move. Are you ready to learn how to keep your machine safe in any situation?

Storage and Transportation Tips

Proper storage and safe transportation are crucial for maintaining your sewing machine's condition and functionality. Whether you're storing your machine for a short period or moving it to a new location, these tips will help ensure its protection.

1. Understanding the Importance of Proper Storage and Transportation

Benefits of proper storage and transportation:
- Prevents dust accumulation and rust formation
- Protects delicate parts from damage
- Maintains alignment and calibration
- Ensures the machine is ready for use when needed

Troubleshooting Tip: Before storing or transporting your machine, always perform a quick check for any loose parts or accessories that might shift during movement.

2. Short-Term Storage Tips

For storage periods of a few days to a few weeks:

a) Cleaning:
- Remove lint and debris from all accessible areas
- Wipe down the exterior with a soft, dry cloth

b) Cover:
- Use a dust cover or a clean pillowcase to protect from dust
- Ensure the cover is breathable to prevent moisture buildup

c) Position:
- Store the machine in its normal operating position
- Avoid placing heavy objects on top of the machine

d) Environment:
- Choose a cool, dry place away from direct sunlight
- Avoid areas with extreme temperature fluctuations

Troubleshooting Tip: If you live in a humid climate, consider using silica gel packets near your stored machine to absorb excess moisture.

3. Long-Term Storage Preparation

For storage periods of several months or more:

a) Thorough Cleaning:
- Perform a deep clean of all parts
- Oil the machine as per manufacturer instructions

b) Needle and Presser Foot:
- Remove the needle to prevent accidental injury
- Lower the presser foot to release tension on the tension discs

c) Bobbin Area:
- Remove the bobbin and bobbin case
- Leave the bobbin area open to prevent moisture accumulation

d) Accessories:
- Store all accessories in a separate, sealed container
- Keep the accessory box with the machine if possible

e) Manual:
- Store the manual with the machine for easy reference

Troubleshooting Tip: Create a "reopening checklist" to remind yourself of all the steps needed to get your machine ready for use after long-term storage.

4. Choosing the Right Storage Location

Ideal storage conditions:
- Stable temperature (around 60-70°F or 15-21°C)
- Low humidity (30-50% relative humidity)
- Away from direct sunlight and heat sources
- Free from dust and pests

5. Transportation Tips

When moving your sewing machine:

a) Original Packaging:
- If possible, use the original box and packing materials
- These are designed to provide optimal protection

b) Alternative Packaging:
- Use a sturdy box slightly larger than the machine
- Wrap the machine in bubble wrap or soft blankets
- Fill empty spaces with packing peanuts or crumpled paper

c) Secure Moving Parts:
- Lock the needle in its highest position
- Secure or remove any detachable parts

d) Carrying the Machine:
- Always lift from the base, not by the handle or arm
- Use both hands for stability

e) In a Vehicle:
• Place the boxed machine on the floor behind a seat
• Avoid the trunk where temperature fluctuations are greatest

Troubleshooting Tip: If transporting in cold weather, allow the machine to come to room temperature before unpacking to prevent condensation.

6. Special Considerations for Different Machine Types

a) Computerized Machines:
• Protect from static electricity during transport
• Consider using anti-static bags for electronic components

b) Sergers/Overlockers:
• Secure or remove the knife mechanism
• Take extra care to protect exposed needles

c) Embroidery Machines:
• Remove and separately pack the embroidery unit
• Secure any movable parts of the embroidery mechanism

7. Reopening After Storage

Before using a machine that's been in storage:

a) Inspection:
• Check for any signs of rust or damage
• Ensure all parts move freely

b) Cleaning:
• Remove any dust that may have accumulated
• Oil the machine if it's been stored for a long time

c) Test Run:
- Thread the machine and sew on scrap fabric
- Check for any unusual noises or performance issues

Troubleshooting Tip: If you notice any musty smells after storage, run the machine in a well-ventilated area for a few minutes to air it out.

8. Creating a Storage/Transportation Kit

Keep these items together for easy machine prep:
- Dust cover
- Silica gel packets
- Small tools for minor adjustments
- Copy of your machine's manual
- List of stored accessories and their locations

9. Dealing with Potential Storage/Transportation Issues

a) Rust Formation:
- Use fine steel wool to gently remove surface rust
- Apply sewing machine oil to prevent further rusting

b) Seized Mechanisms:
- Apply lubricant and gently work the mechanism
- Seek professional help if parts remain stuck

c) Misalignment:
- Perform a thorough calibration check
- Consult a technician if serious misalignment is suspected

10. Insurance and Documentation

- Consider insuring valuable machines during storage or transport
- Take photos of your machine and serial number for documentation
- Keep receipts and warranty information with the machine

Remember, proper storage and careful transportation are key to maintaining your sewing machine's longevity and performance. By following these guidelines, you'll ensure that your machine remains in top condition, ready to tackle your next project whenever you need it.

Developing good storage and transportation habits will not only protect your investment but also give you peace of mind, knowing that your trusty sewing companion is safe and well-cared for.

In our next section, we'll explore troubleshooting guides for common sewing machine issues, empowering you to diagnose and solve problems quickly and effectively. Are you ready to become a sewing machine detective?

Chapter 9
Troubleshooting Guide
Diagnosing Common Issues

Being able to diagnose common sewing machine issues can save you time, money, and frustration. Let's explore how to identify and address the most frequent problems.

1. The Importance of Systematic Troubleshooting

Benefits of a methodical approach:
- Quickly identifies the root cause of problems
- Prevents unnecessary disassembly
- Saves time and reduces frustration
- Helps avoid causing additional damage

Troubleshooting Tip: Always start with the simplest possible cause and work your way to more complex issues. Often, the problem has a simple solution!

2. Common Issue: Machine Won't Turn On

Possible Causes:
a) Power Supply Problems:
   - Check if the machine is plugged in
   - Test the outlet with another device
   - Inspect the power cord for damage

b) Faulty On/Off Switch:
   - Check if the switch moves freely
   - Listen for a clicking sound when switched

c) Internal Wiring Issues:
  • Look for any visible damage or loose connections

Action Steps:
1. Ensure proper connection to power source
2. Try a different outlet
3. Examine power cord and replace if damaged
4. If problems persist, consult a technician for internal checks

3. Common Issue: Thread Breaking

Possible Causes:
a) Incorrect Threading:
  • Rethread the machine, ensuring proper path

b) Tension Problems:
  • Check and adjust upper and lower thread tensions

c) Damaged Needle or Wrong Size:
  • Replace with appropriate needle for fabric

d) Poor Quality or Old Thread:
  • Use high-quality, new thread

e) Rough Spots in Thread Path:
  • Check for burrs on needle, throat plate, or tension discs

Action Steps:
1. Completely rethread the machine
2. Adjust tension settings
3. Replace needle and/or thread
4. Clean and check for any rough spots in thread path

Troubleshooting Tip: When dealing with thread breakage, always start by rethreading both the upper thread and bobbin. This simple step often resolves the issue.

4. Common Issue: Skipped Stitches

Possible Causes:
a) Bent or Dull Needle:
- Replace with a new, appropriate needle

b) Incorrect Needle Size for Fabric:
- Use the right needle type and size for your material

c) Improper Threading:
- Ensure correct threading path

d) Incorrect Presser Foot Pressure:
- Adjust pressure for fabric thickness

Action Steps:
1. Replace the needle
2. Verify correct needle size and type for fabric
3. Rethread the machine
4. Adjust presser foot pressure

5. Common Issue: Uneven Stitches

Possible Causes:
a) Unbalanced Thread Tension:
- Adjust upper and lower tensions

b) Inconsistent Fabric Feeding:
- Check feed dogs for lint buildup or damage

c) Incorrect Presser Foot:
- Use appropriate foot for the stitch type

d) Uneven Pressure on Fabric:
- Ensure even guidance of fabric while sewing

Action Steps:
1. Balance thread tensions
2. Clean or repair feed dogs
3. Verify correct presser foot
4. Practice guiding fabric evenly

Troubleshooting Tip: When adjusting tension for uneven stitches, make small, incremental changes and test on scrap fabric after each adjustment.

6. Common Issue: Machine Making Unusual Noises

Possible Causes:
a) Need for Oiling:
- Lubricate as per machine manual

b) Lint or Thread Jam:
- Clean bobbin area and feed dogs

c) Bent Needle Hitting Parts:
- Replace needle and check alignment

d) Loose or Worn Parts:
- Tighten visible screws, check for wear

Action Steps:
1. Oil the machine
2. Perform a thorough cleaning
3. Replace the needle

4. Tighten loose parts, seek professional help for internal issues

7. Common Issue: Fabric Not Feeding Properly

Possible Causes:
a) Feed Dogs Lowered or Clogged:
- Raise feed dogs, clean thoroughly

b) Incorrect Presser Foot Pressure:
- Adjust for fabric thickness

c) Stitch Length Set to Zero:
- Check and adjust stitch length

d) Worn Feed Dogs:
- Inspect for wear, consider replacement

Action Steps:
1. Check feed dog position and clean
2. Adjust presser foot pressure
3. Verify stitch length setting
4. Inspect feed dogs for wear

Troubleshooting Tip: For very light or slippery fabrics, try using a walking foot to improve feeding.

8. Common Issue: Needle Breaking

Possible Causes:
a) Bent Needle:
- Replace with new, straight needle

b) Incorrect Needle Insertion:
- Ensure needle is fully inserted and properly oriented

c) Pulling Fabric While Sewing:
- Let the feed dogs guide the fabric

d) Wrong Needle Size for Fabric:
- Use appropriate needle for material thickness

Action Steps:
1. Replace needle
2. Check needle insertion
3. Practice proper fabric guidance
4. Verify correct needle size for fabric

9. General Troubleshooting Approach

When encountering any issue:
1. Identify the specific problem (e.g., sound, stitch quality)
2. Consider recent changes (new thread, fabric, or settings)
3. Check the obvious (threading, power, needle condition)
4. Consult your machine's manual for model-specific advice
5. Test solutions on scrap fabric
6. If problems persist, seek professional help

Troubleshooting Tip: Keep a log of issues and solutions. This personal troubleshooting guide can be invaluable for recurring problems.

10. When to Seek Professional Help

Consider professional servicing if:
- You've tried basic troubleshooting without success
- The issue involves internal mechanisms you're not comfortable accessing
- You notice electrical problems or burning smells
- The machine is under warranty

Remember, effective troubleshooting is a skill that improves with practice. Don't get discouraged if you can't solve every problem immediately. With time and experience, you'll become more adept at diagnosing and resolving issues, keeping your sewing machine running smoothly.

In our next section, we'll explore step-by-step problem-solving flowcharts to further enhance your troubleshooting skills. Are you ready to become a sewing machine problem-solving expert?

Step-by-Step Problem-Solving Flowcharts

Problem-solving flowcharts provide a structured approach to troubleshooting, guiding you through a series of questions and actions to identify and resolve issues efficiently.

1. Understanding Flowcharts

A flowchart is a diagram that represents a process, showing steps as boxes of various kinds, and their order by connecting them with arrows. For sewing machine troubleshooting, flowcharts can help you:
- Systematically approach problems
- Consider all possible causes
- Take appropriate actions in a logical order

Troubleshooting Tip: When using flowcharts, always start at the beginning and follow each step, even if you think you know the cause. This ensures you don't overlook any potential issues.

2. Basic Flowchart Symbols

- Oval: Start or end of the process
- Rectangle: Action or operation
- Diamond: Decision point (usually a yes/no question)
- Arrow: Direction of flow

3. Example Flowchart: Thread Breaking Issue

Let's create a flowchart for a common problem: thread breaking.

Start → Is the machine threaded correctly?
↓ es ↓ No
↓ Rethread machine → Test → Problem solved?
↓ ↓ No ↓ Yes
↓ ↓ End
Is the needle inserted correctly and undamaged?
↓ Yes ↓ No
↓ Replace needle → Test → Problem solved?
↓ ↓ No ↓ Yes
↓ ↓ End
Check tension → Adjust if necessary → Test → Problem solved?
↓ No ↓ Yes
↓ End
Is the thread old or poor quality?
↓ Yes ↓ No
↓ Seek professional help
Replace thread → Test → Problem solved?
↓ No ↓ Yes
↓ End
Seek professional help

4. Creating Your Own Flowcharts

To create effective troubleshooting flowcharts:
a) Identify the main problem
b) List all possible causes
c) Arrange causes from most common/simple to least common/complex
d) Add decision points and action steps
e) Include test steps after each action
f) End with either problem resolution or seeking professional help

Troubleshooting Tip: Use different colors for decision points, actions, and endpoints to make your flowchart easier to follow.

5. Flowchart for Skipped Stitches

Start → Is the needle bent or dull?
↓ No ↓ Yes
↓ Replace needle → Test → Problem solved?
↓↓ No ↓ Yes
↓↓ End
Is the needle size correct for the fabric?
↓ Yes ↓ No
↓ Change to correct needle size → Test → Problem solved?
↓↓ No ↓ Yes
↓↓ End
Is the machine threaded correctly?
↓ Yes ↓ No
↓ Rethread machine → Test → Problem solved?
↓↓ No ↓ Yes
↓↓ End
Check bobbin installation and tension
↓
Adjust if necessary → Test → Problem solved?
↓ No ↓ Yes
↓ End
Seek professional help

6. Flowchart for Machine Not Turning On

Start → Is the machine plugged in?
↓ Yes ↓ No
↓ Plug in machine → Test → Problem solved?
↓↓ No ↓ Yes

↓↓ End
Is the outlet working?
↓ Yes ↓ No
↓ Try different outlet → Test → Problem solved?
↓↓ No ↓ Yes
↓↓ End
Check power cord for damage
↓ No damage ↓ Damage found
↓ Replace cord → Test → Problem solved?
↓↓ No ↓ Yes
↓↓ End
Check on/off switch
↓
If faulty, seek professional help

7. Tips for Using Flowcharts Effectively

- Start at the top and follow each step sequentially
- Perform each test thoroughly before moving on
- Keep track of where you are in the flowchart
- Document your findings at each step

8. Customizing Flowcharts for Your Machine

- Consult your machine's manual for model-specific issues
- Add steps relevant to your machine's features
- Include common problems you've encountered

Troubleshooting Tip: Create a master flowchart for general issues, and separate, detailed flowcharts for complex problems specific to your machine model.

9. Digital Flowchart Tools

Consider using digital tools to create and store your flowcharts:
- Microsoft Visio
- Lucidchart
- Draw.io
- SmartDraw

These tools allow for easy updates and sharing of your flowcharts.

10. Combining Flowcharts with Maintenance Logs

- Reference your maintenance log when using flowcharts
- Note which flowchart steps resolved issues in your log
- Use this information to predict and prevent future problems

11. Teaching Others to Use Your Flowcharts

If others use your sewing machine:
- Explain the basic structure of the flowcharts
- Encourage them to follow the steps precisely
- Make flowcharts easily accessible near the machine

Troubleshooting Tip: Laminate printed flowcharts or keep them in clear plastic sleeves to protect them from wear and tear in your sewing area.

By creating and using problem-solving flowcharts, you'll develop a systematic approach to troubleshooting your sewing machine. This not only helps you resolve issues more quickly but also deepens your understanding of how your machine works.

Remember, flowcharts are living documents. As you gain more experience with your machine, don't hesitate to update and refine your flowcharts to make them even more effective.

In our next section, we'll discuss when it's appropriate to seek professional help for your sewing machine issues. Are you ready to learn when DIY troubleshooting reaches its limits?Certainly! Let's dive into creating and using Step-by-Step Problem-Solving Flowcharts for sewing machine issues. These visual guides can be incredibly helpful in systematically diagnosing and resolving problems.

When to Seek Professional Help

While many sewing machine issues can be resolved at home, there are times when professional assistance is necessary. Recognizing these situations can save you time, prevent further damage, and ensure your machine receives the care it needs.

1. Understanding the Importance of Professional Service

Benefits of seeking professional help:
- Access to specialized tools and parts
- Expert diagnosis of complex issues
- Prevention of accidental damage from DIY attempts
- Maintenance of warranty validity

Troubleshooting Tip: Establish a relationship with a reputable sewing machine technician before you have a major issue. This can be invaluable when you need urgent help.

2. Signs It's Time for Professional Help

a) Persistent Issues:
- Problems that recur despite your best troubleshooting efforts
- Issues that worsen over time despite maintenance

b) Unusual Noises:
- Grinding, clicking, or loud buzzing that doesn't resolve with oiling
- Sudden changes in the machine's normal operating sound

c) Electrical Problems:
- Inconsistent power or complete failure to turn on
- Burning smells or visible sparks

d) Timing Issues:
- Persistent skipped stitches that don't resolve with needle changes or re-threading
- Hook and needle synchronization problems

e) Physical Damage:
- Visible cracks in the body or essential components
- Bent or misaligned parts that affect operation

3. Scenarios Requiring Professional Intervention

a) After Accidental Damage:
- Machine was dropped or had heavy objects fall on it
- Liquid spills inside the machine

b) Computerized Machine Malfunctions:
- Error codes that persist after basic troubleshooting
- Screen or software issues

c) Tension Assembly Problems:
- Unable to achieve proper tension despite adjustments
- Tension discs not moving freely

d) Feed Dog Issues:
- Feed dogs not moving or moving erratically
- Fabric feeding inconsistently despite adjustments

Troubleshooting Tip: If you're unsure whether an issue requires professional help, consult your machine's manual. Many manuals have a section on when to seek professional service.

4. When DIY Repairs Reach Their Limit

Consider professional help when:
- You've exhausted all troubleshooting steps in your manual
- The repair requires specialized tools you don't have
- You're uncomfortable disassembling complex parts
- The issue involves internal electronic components

5. Warranty Considerations

- Check your warranty terms before attempting any repairs
- Some warranties are voided by unauthorized DIY repairs
- If under warranty, always consult the manufacturer or authorized dealer first

6. Cost-Benefit Analysis

Before deciding on professional repair, consider:
- The age and value of your machine
- The cost of repair versus replacement
- The complexity and time involved in DIY attempts

Troubleshooting Tip: Keep a record of repair costs. If they start to approach the cost of a new machine, it might be time to consider replacement.

7. Preparing for Professional Service

When you decide to seek professional help:
- Clean your machine thoroughly
- Gather all accessories and attachments
- Prepare a detailed description of the issue
- Bring samples of problematic stitching if applicable

8. Choosing the Right Professional

Look for a technician who:
- Is certified or authorized by your machine's manufacturer
- Has experience with your specific machine model
- Offers a warranty on their work
- Has positive reviews or recommendations

9. Questions to Ask a Professional

- What's the estimated cost and timeframe for the repair?
- Are there any preventive measures to avoid this issue in the future?
- Can they perform a general service while repairing the specific issue?

10. Learning from Professional Repairs

- Ask the technician to explain the issue and repair process
- Inquire about maintenance tips to prevent future problems
- Request a detailed invoice explaining all work done

Troubleshooting Tip: Use the professional repair as a learning opportunity. Understanding what went wrong can help you prevent similar issues in the future.

11. When Replacement Might Be Better Than Repair

Consider replacing your machine if:
- Repair costs exceed 50% of the cost of a new, comparable machine
- Parts are no longer available for your model
- The machine has had multiple major repairs in a short time
- Upgrading would significantly improve your sewing capabilities

12. Preventive Professional Maintenance

- Schedule regular professional servicing (e.g., annually)
- This can catch potential issues before they become major problems
- Professional tuning can optimize your machine's performance

13. After Professional Service

- Test the machine thoroughly after getting it back
- Follow any new maintenance instructions provided
- Update your maintenance log with details of the professional service

Remember, seeking professional help when needed is a sign of responsible machine ownership, not a failure in your ability to maintain your sewing machine. Professional technicians have the expertise, tools, and experience to handle complex issues safely and effectively.

By knowing when to seek professional help, you're ensuring the longevity and optimal performance of your sewing machine. This knowledge, combined with your own maintenance efforts, will keep your machine running smoothly for years to come.

In our next section, we'll explore resources and references to further enhance your sewing machine maintenance and repair knowledge. Are you ready to expand your sewing machine expertise?

Chapter 10
Resources and References
Recommended Tools and Supplies

Having the right tools and supplies on hand is crucial for effective sewing machine maintenance and troubleshooting. Let's explore the must-have items for your sewing machine toolkit.

1. Basic Tool Kit Essentials

a) Screwdrivers:
- Small flathead and Phillips head screwdrivers
- Precision screwdriver set for tiny screws

b) Pliers:
- Needle-nose pliers for reaching tight spaces
- Regular pliers for general use

c) Tweezers:
- Fine-tipped for precision work
- Curved tweezers for hard-to-reach areas

d) Brushes:
- Lint brush for cleaning debris
- Small, soft paintbrush for delicate areas

e) Flashlight:
- Preferably LED for bright, focused light
- Consider a headlamp for hands-free operation

2. Cleaning Supplies

a) Cleaning Cloths:
- Lint-free microfiber cloths
- Cotton swabs for small spaces

b) Compressed Air:
- Canned air for blowing out lint and debris
- Use cautiously to avoid pushing debris further into the machine

c) Cleaning Solutions:
- Isopropyl alcohol for removing stubborn grime
- Specialized sewing machine cleaner (if recommended by manufacturer)

d) Vacuum Attachments:
- Mini vacuum attachments for thorough cleaning

Troubleshooting Tip: Always unplug your machine before cleaning. Be cautious with liquids around electronic components.

3. Lubrication and Maintenance Supplies

a) Sewing Machine Oil:
- High-quality, clear sewing machine oil
- Long-nozzled oiler for precise application

b) Grease:
- Sewing machine grease for gears (if required by your model)

c) Lubricant Applicators:
- Syringe or pipette for precise oil application
- Lint-free cloth for wiping excess oil

Troubleshooting Tip: Use only sewing machine-specific oils and lubricants. Avoid household oils or WD-40, which can damage your machine.

4. Replacement Parts

a) Needles:
- Assortment of needle types and sizes
- Universal needles for general use

b) Bobbins:
- Extra bobbins compatible with your machine
- Pre-wound bobbins for convenience

c) Light Bulbs:
- Replacement bulbs specific to your machine model

d) Belts:
- Spare drive belt (if your machine uses one)

Troubleshooting Tip: Always keep spare needles and bobbins on hand. A bent needle or empty bobbin can halt your project unexpectedly.

5. Diagnostic Tools

a) Multimeter:
- For testing electrical components
- Choose one with both voltage and continuity testing

b) Tension Gauge:
- For accurately measuring and adjusting thread tension

c) Timing Gauge:
- Helps check and adjust hook timing

d) Magnifying Glass:
- For inspecting small parts and stitches

Troubleshooting Tip: Invest in quality tools. They'll last longer and reduce the risk of stripping screws or damaging delicate parts.Troubleshooting Tip: Learn to use your diagnostic tools properly. Incorrect use can lead to misdiagnosis of issues.

6. Safety Equipment

a) Safety Glasses:
- Protect eyes from flying debris or broken needles

b) Work Gloves:
- Thin, close-fitting gloves for protection during repairs

c) Anti-static Wrist Strap:
- For working on computerized machines to prevent static damage

7. Organizational Tools

a) Magnetic Parts Tray:
- Keeps small metal parts from getting lost

b) Compartment Organizer:
- For storing various screws, needles, and small parts

c) Label Maker:
- For labeling parts and settings during disassembly

Troubleshooting Tip: Take photos or videos during disassembly to aid in reassembly. Label parts and note their positions.

8. Specialty Tools

a) Needle Threader:
- Helps with difficult threading, especially in older machines

b) Seam Ripper:
- For removing stitches during testing or troubleshooting

c) Thread Stand:
- External thread stand for large spools or specialty threads

d) Bobbin Winder:
- External bobbin winder for convenience

9. Reference Materials

a) Machine Manual:
- Keep your machine's manual easily accessible

b) Maintenance Log:
- Notebook or digital app for tracking maintenance and repairs

c) Troubleshooting Flowcharts:
- Create or obtain flowcharts for common issues

Troubleshooting Tip: Create a digital copy of your machine's manual in case the physical copy is lost or damaged.

10. Advanced Repair Tools (for experienced users)

a) Soldering Iron:
- For minor electrical repairs

b) Hook Adjustment Gauge:
- For precise timing adjustments

c) Feeler Gauges:
- For measuring small clearances

11. Building Your Toolkit

- Start with basic essentials and add specialized tools as needed
- Consider purchasing a pre-assembled sewing machine repair kit
- Gradually upgrade tools as your skills improve

12. Storing Your Tools and Supplies

- Use a dedicated toolbox or organizer for sewing machine tools
- Keep tools clean and dry to prevent rust
- Store lubricants and cleaning solutions properly to prevent leaks

Troubleshooting Tip: Regularly check and maintain your tools. A well-maintained toolkit ensures you're always ready for machine maintenance.

13. Where to Find Tools and Supplies

- Sewing machine dealers and repair shops
- Online retailers specializing in sewing supplies
- General hardware stores for basic tools

Remember, having the right tools and supplies is half the battle in effective sewing machine maintenance and repair. A well-equipped toolkit allows you to handle routine maintenance confidently and address minor issues promptly.

As you become more familiar with your machine and more confident in your repair skills, you may find yourself gradually expanding your toolkit. Always prioritize quality when selecting tools, as they are an investment in your sewing machine's longevity.

In our next section, we'll explore parts sourcing guides to help you find the right components for your sewing machine repairs. Are you ready to become a master at locating the perfect parts for your machine?

Parts Sourcing Guide

Finding the correct parts for your sewing machine is crucial for successful repairs and maintenance. This guide will help you locate and purchase the right components efficiently and effectively.

1. Understanding the Importance of Correct Parts

Using the right parts ensures:
- Proper fit and function
- Maintenance of machine performance
- Prevention of further damage
- Preservation of warranty (if applicable)

Troubleshooting Tip: Always double-check part numbers and machine compatibility before ordering. A slight variation in model number can lead to incompatible parts.

2. Identifying Your Machine and Part Needs

Before sourcing parts:
a) Locate your machine's model number and serial number
b) Identify the specific part needed (consult your manual or a parts diagram)
c) Note any alternative part numbers or compatible substitutes

3. Common Sewing Machine Parts to Source

- Needles and needle plates
- Bobbins and bobbin cases
- Presser feet
- Drive belts

- Light bulbs
- Tension assembly components
- Feed dogs

4. Sources for Sewing Machine Parts

a) Authorized Dealers:
- Often have direct access to manufacturer parts
- Can provide guidance on compatibility
- May be more expensive but offer guaranteed authenticity

b) Online Retailers:
- Websites specializing in sewing machine parts
- General marketplaces (e.g., Amazon, eBay)
- Often have a wide selection and competitive prices

c) Sewing Machine Repair Shops:
- May stock common parts
- Can order specific parts for you
- Offer expertise in part selection

d) Manufacturer Websites:
- Direct source for original parts
- May offer parts lookup tools

e) Salvage and Second-hand Sources:
- For rare or discontinued parts
- Requires caution and thorough inspection before use

Troubleshooting Tip: When sourcing from second-hand sellers, ask for clear photos and detailed descriptions to ensure part compatibility and condition.

5. Strategies for Efficient Parts Sourcing

a) Use Online Parts Diagrams:
- Many manufacturers provide interactive parts diagrams
- Helps in identifying exact part names and numbers

b) Join Sewing Machine Forums:
- Community members often share sourcing tips
- Can help identify alternative parts for discontinued models

c) Create a Parts Inventory:
- Keep a list of commonly needed parts
- Note part numbers and potential sources

d) Establish Relationships with Suppliers:
- Regular customers often receive better service and deals
- Suppliers may alert you to sales or hard-to-find parts

6. Dealing with Discontinued Parts

When a part is no longer manufactured:
a) Search for new old stock (NOS)
b) Consider compatible parts from newer models
c) Explore 3D printing options for simple plastic parts
d) Consult with repair technicians for custom solutions

Troubleshooting Tip: For vintage machines, join collector groups. They often have valuable information on sourcing rare parts.

7. Understanding Part Compatibility

- OEM (Original Equipment Manufacturer) vs. aftermarket parts
- Universal parts vs. machine-specific components
- Upgraded parts that replace discontinued ones

8. Evaluating Part Quality

When sourcing, consider:
- Material quality (especially for moving parts)
- Reviews and ratings from other users
- Warranty or return policy
- Price in relation to OEM parts

9. Bulk Buying vs. As-Needed Purchasing

Pros of bulk buying:
- Cost savings on frequently used parts
- Immediate availability when needed

Cons of bulk buying:
- Initial higher cost
- Storage requirements
- Risk of parts becoming obsolete

Troubleshooting Tip: For parts that wear out regularly (like needles or bobbins), bulk buying can be economical. For rare repairs, as-needed purchasing is often better.

10. International Sourcing Considerations

When sourcing internationally:
- Be aware of potential import duties
- Consider shipping times and costs
- Check voltage compatibility for electrical components
- Understand return policies for international orders

11. Emergency Sourcing Techniques

For urgent repairs:
- Contact local sewing groups for spare parts
- Check if repair shops offer loaner parts
- Consider temporary fixes with universal parts until the correct part arrives

12. Maintaining a Parts Sourcing Log

Keep track of:
- Part numbers and descriptions
- Successful and unsuccessful sources
- Prices paid and delivery times
- Quality and fit of received parts

13. Ethical Considerations in Parts Sourcing

- Support authorized dealers when possible
- Be cautious of counterfeit parts that may damage your machine
- Consider the environmental impact of shipping and packaging
- Gears and cam stacks

14. Future-Proofing Your Parts Supply

- For beloved or critical machines, consider purchasing spare parts proactively
- Stay informed about manufacturer plans for parts availability
- Learn about alternative or adaptable parts for your machine model

Troubleshooting Tip: If you find a reliable source for hard-to-find parts, bookmark it and consider sharing with the sewing community (while ensuring you don't exhaust the supply for your own needs).

Remember, successful parts sourcing is a combination of research, patience, and sometimes creativity. By developing good sourcing habits and maintaining detailed records, you'll be well-equipped to keep your sewing machine in top condition, no matter its age or model.

Effective parts sourcing ensures that you can maintain and repair your sewing machine efficiently, extending its life and maintaining its performance. With practice, you'll become adept at finding the right parts quickly and cost-effectively.

In our next section, we'll explore online communities and further learning resources to continually enhance your sewing machine maintenance and repair skills. Are you ready to connect with fellow enthusiasts and expand your knowledge?

Online Communities and Further Learning

In the digital age, a wealth of information and support is available online for those interested in sewing machine maintenance and repair. Let's dive into the various resources you can utilize to enhance your skills and connect with like-minded individuals.

1. Online Forums and Discussion Boards

Popular sewing machine forums include:
• PatternReview.com's Machine Embroidery and Sewing Machine forum
• Quiltingboard.com's machine repair section
• Reddit's r/sewing and r/vintagesewing communities

Benefits of participating in forums:
• Access to collective knowledge and experience
• Quick responses to specific questions
• Exposure to diverse machine models and issues

Troubleshooting Tip: When posting a question, provide as much detail as possible about your machine model and the issue you're facing. Clear, well-lit photos can be extremely helpful.

2. Social Media Groups

Facebook Groups:
• "Sewing Machine Repair and Maintenance"
• "Vintage Sewing Machine Enthusiasts"
• Brand-specific groups (e.g., "Singer Featherweight Owners")

Instagram:
- Follow hashtags like #sewingmachinerepair or #vintagesewingmachines
- Connect with sewing machine technicians and enthusiasts

Benefits of social media engagement:
- Real-time interactions and quick tips
- Visual guides and tutorial videos
- Networking with experts and fellow enthusiasts

3. YouTube Channels and Video Tutorials

Popular channels for sewing machine repair:
- "The Treasure Cellar" for vintage machine repairs
- "Man Sewing" for general sewing machine maintenance
- "Sewing Parts Online" for specific repair tutorials

How to maximize learning from video content:
- Create playlists for different types of repairs or machine models
- Practice alongside the videos, pausing when needed
- Engage with creators through comments for clarifications

Troubleshooting Tip: When following a video tutorial, always verify that the instructions are suitable for your specific machine model to avoid potential damage.

4. Online Courses and Webinars

Platforms offering sewing machine repair courses:
- Udemy
- Craftsy
- Textile Courses Online

Benefits of structured online learning:
- Comprehensive coverage of topics
- Often includes quizzes and assignments for skill reinforcement
- Certificates of completion (useful for those considering professional repair work)

5. Manufacturer Resources

Many sewing machine manufacturers offer:
- Online manuals and parts catalogs
- Troubleshooting guides
- Video tutorials for machine-specific maintenance

How to access these resources:
- Check the official website of your machine's manufacturer
- Register your product for access to exclusive content
- Subscribe to manufacturer newsletters for updates and tips

6. E-books and Digital Publications

Look for e-books on topics like:
- Vintage sewing machine restoration
- Advanced troubleshooting techniques
- Specific machine model repair guides

Where to find them:
- Amazon Kindle store
- Specialized sewing and crafting websites
- Direct from sewing machine repair experts

Troubleshooting Tip: When using e-books for guidance, cross-reference information with your machine's manual to ensure compatibility of techniques.

7. Podcasts

Sewing and machine repair podcasts:
• "Sew & Tell" by BERNINA
• "The Self Sewn Wardrobe" (occasionally covers machine topics)

Benefits of podcast learning:
• Learn while multitasking
• Stay updated on latest trends and technologies
• Hear from industry experts and experienced repairers

8. Virtual Workshops and Live Streams

Many experts offer:
• Live Q&A sessions
• Virtual repair demonstrations
• Interactive troubleshooting workshops

How to find these opportunities:
• Follow sewing machine technicians and educators on social media
• Check sewing machine dealer websites for event listings
• Join mailing lists of sewing education platforms

9. Online Sewing Machine Museums

Virtual museums offer:
• Detailed information on vintage and antique machines
• Historical context for machine development
• Sometimes include repair and restoration tips

Examples:
- The International Sewing Machine Collectors' Society (ISMACS) website
- Various collector websites dedicated to specific brands

10. Collaborative Projects and Challenges

Participate in:
- Group restoration projects of vintage machines
- Troubleshooting challenges where community members solve issues together
- Machine modification or customization projects

Benefits:
- Hands-on learning in a supportive environment
- Exposure to diverse machine types and problems
- Building a network of fellow enthusiasts

11. Creating Your Own Online Content

Consider:
- Starting a blog about your repair experiences
- Creating video tutorials for specific repairs you've mastered
- Sharing before-and-after photos of your projects on social media

Benefits of content creation:
- Reinforces your own learning
- Connects you with others in the community
- Potentially establishes you as an expert in the field

Troubleshooting Tip: When creating content, always emphasize safety precautions and remind viewers to consult their machine's manual.

12. Staying Updated with Industry News

Follow:
- Sewing machine manufacturer blogs
- Sewing technology news websites
- Trade publications related to sewing and textile industries

Why it's important:
- Learn about new technologies and repair techniques
- Stay informed about parts availability and machine longevity
- Understand trends that might affect future repair needs

Remember, the world of online learning and community engagement for sewing machine enthusiasts is vast and ever-growing. By actively participating in these communities and utilizing various learning resources, you'll continually expand your knowledge and skills.

Engaging with online communities not only enhances your own abilities but also allows you to contribute to the collective knowledge of the sewing machine repair community. Whether you're a hobbyist or considering a professional path in sewing machine repair, these resources can be invaluable in your journey.

As you explore these online resources, always approach the information critically, verify techniques with reliable sources, and never hesitate to ask questions when in doubt. The sewing machine community is generally very supportive and eager to help fellow enthusiasts grow their skills.

Appendices
Sewing Machine Repair Glossary

This glossary aims to demystify the technical terms you might encounter during sewing machine repair and maintenance. Understanding these terms will help you communicate more effectively about machine issues and follow repair instructions with greater ease.

1. Anatomy of a Sewing Machine

- Arm: The horizontal upper part of the machine that holds the needle bar mechanism.
- Bed: The flat, horizontal surface where fabric rests during sewing.
- Free Arm: A narrower extension of the bed, useful for sewing tubular items.
- Throat Plate (Needle Plate): The metal plate with holes through which the needle passes.

Troubleshooting Tip: When diagnosing issues, always start by identifying which part of the machine is involved. This can quickly narrow down potential problems.

2. Stitch Formation Components

- Bobbin: A small spool that holds the lower thread.
- Bobbin Case: Holds the bobbin and controls lower thread tension.
- Hook: Catches the upper thread to form a loop around the bobbin thread.
- Take-Up Lever: Pulls thread from the spool and controls thread slack.

3. Fabric Feed Mechanism

• Feed Dogs: Toothed metal bars that move fabric through the machine.
• Presser Foot: Holds fabric in place against the feed dogs.
• Differential Feed: In sergers, allows for adjusting fabric feed rates to prevent stretching or puckering.

Troubleshooting Tip: If you're experiencing uneven stitches, check the feed dogs for lint buildup or damage.

4. Tension-Related Terms

• Tension Discs: Control the pressure on the upper thread.
• Tension Spring: Part of the tension assembly that applies pressure to the thread.
• Tension Regulator: Allows adjustment of upper thread tension.

5. Needle and Thread Path

• Thread Guide: Directs thread from the spool to the needle.
• Needle Bar: Holds the needle and moves it up and down.
• Needle Clamp: Secures the needle in place.
• Eye of the Needle: The hole in the needle through which thread passes.

6. Motor and Power Transmission

• Drive Belt: Transfers power from the motor to the main shaft.
• Handwheel: Allows manual control of the needle position.
• Motor: Provides power to operate the machine.
• Clutch: Disengages the needle for bobbin winding.

Troubleshooting Tip: A slipping drive belt can cause inconsistent stitching. Check belt tension if you notice this issue.

7. Lubrication and Maintenance Terms

• Gib Hook: The hook component in some machines that requires periodic adjustment.
• Race: The track in which the hook assembly rotates.
• Sewing Machine Oil: Specialized oil for lubricating machine parts.

8. Stitch Types and Adjustments

• Stitch Length: The distance between individual stitches.
• Stitch Width: The lateral distance a needle travels in stitches like zigzag.
• Straight Stitch: A basic stitch moving only forward.
• Zigzag Stitch: A stitch that moves side to side as well as forward.

9. Specialized Machine Parts

• Cam: A shaped wheel that controls stitch patterns in mechanical machines.
• Needle Position Lever: Allows changing the needle position left to right.
• Reverse Lever: Enables backward stitching.

10. Electrical Components

• Foot Controller (Pedal): Controls the machine's speed.
• LED Light: Illuminates the sewing area in modern machines.

- Rheostat: Controls motor speed in the foot pedal.

Troubleshooting Tip: If your machine won't turn on, always check the foot controller connection first before assuming more serious electrical issues.

11. Serger-Specific Terms

- Loopers: Form stitches in conjunction with needles in sergers.
- Cutting Width: The distance between the needle and the cutting blade in a serger.
- Stitch Finger: Guides the thread for proper looping in sergers.

12. Computerized Machine Terms

- LCD Screen: Displays stitch selections and machine settings.
- Embroidery Unit: Attachment for computerized embroidery.
- USB Port: Allows uploading of custom designs in some machines.

13. Repair and Diagnostic Terms

- Timing: The synchronization between the needle and hook movements.
- Calibration: Adjusting machine settings for optimal performance.
- Burr: A rough edge on metal parts that can damage thread or fabric.

Troubleshooting Tip: If you suspect timing issues, always consult your machine's manual for the correct procedure, as timing can vary between models.

14. Tools and Supplies

- Screwdriver: Used for adjusting and disassembling machine parts.
- Tweezers: Helpful for removing thread jams or debris in tight spaces.
- Multimeter: Used for testing electrical components.

15. Common Issues

- Bird Nesting: Tangled mess of thread on the underside of fabric.
- Skipped Stitches: When the machine fails to form a stitch at regular intervals.
- Thread Bunching: Loops of thread forming on top or bottom of fabric.

Troubleshooting Tip: When encountering any of these common issues, always start by rethreading both the upper thread and bobbin, as improper threading is a frequent cause.

This glossary serves as a foundation for understanding sewing machine terminology. As you delve deeper into repairs and maintenance, you'll likely encounter more specialized terms. Don't hesitate to ask for clarification in online communities or consult machine-specific manuals when you come across unfamiliar terms.

Remember, understanding the language of sewing machine repair is key to effective troubleshooting and maintenance. Keep this glossary handy as a quick reference during your repair projects, and consider adding to it as you learn new terms and concepts in your sewing machine repair journey.

Brand-Specific Maintenance Tips

Different sewing machine brands often have unique features and requirements. Understanding these can help you maintain your machine more effectively and troubleshoot brand-specific issues.

1. Singer Sewing Machines

Singer is one of the oldest and most well-known sewing machine brands.

Key Maintenance Tips:
- Oil frequently, especially for vintage models
- Clean the feed dogs and bobbin area after every few projects
- Check the belt tension regularly on older models

Troubleshooting Tip for Singer Machines: If you experience timing issues on older Singer models, check the position of the balance wheel set screw. It should align with the indentation on the shaft.

2. Brother Sewing Machines

Brother machines are known for their user-friendly features and reliability.

Key Maintenance Tips:
- Use only Brother-approved bobbins to prevent jams
- Clean the race hook area regularly
- Update firmware on computerized models

Troubleshooting Tip for Brother Machines: If you encounter error codes, consult the manual for specific meanings. Many Brother machines have self-diagnostic features.

3. Janome Sewing Machines

Janome machines are praised for their smooth operation and durability.

Key Maintenance Tips:
- Use high-quality thread to prevent lint buildup
- Clean and oil the hook race area regularly
- Check and tighten presser foot screws periodically

Troubleshooting Tip for Janome Machines: If experiencing skipped stitches, ensure you're using the correct needle type and size for your fabric.

4. Bernina Sewing Machines

Bernina is known for high-end, precision machines.

Key Maintenance Tips:
- Use only Bernina-approved accessories and parts
- Clean the thread cutter mechanism regularly
- Schedule professional servicing annually for optimal performance

Troubleshooting Tip for Bernina Machines: If the machine is sluggish, check for lint in the upper thread guide area, a often-overlooked spot in Bernina machines.

5. Pfaff Sewing Machines

Pfaff machines are known for their innovative IDT (Integrated Dual Feed) system.

Key Maintenance Tips:
- Clean and lubricate the IDT system regularly
- Use high-quality, lint-free thread to prevent buildup
- Check and clean the thread sensor on computerized models

Troubleshooting Tip for Pfaff Machines: If the IDT system isn't engaging properly, check for thread or lint caught in the mechanism.

6. Husqvarna Viking Sewing Machines

Viking machines are known for their durability and advanced features.

Key Maintenance Tips:
- Clean the sensor system in computerized models regularly
- Use only recommended needle types to prevent damage
- Keep the bobbin area free from lint and thread pieces

Troubleshooting Tip for Viking Machines: If experiencing tension issues, check that the threads are properly seated in all guides, including the pre-tension guide.

7. Juki Sewing Machines

Juki is renowned for industrial-grade machines but also offers domestic models.

Key Maintenance Tips:
• Oil more frequently, especially for industrial models
• Clean the bobbin case area after each day of use
• Check and adjust belt tension regularly on industrial models

Troubleshooting Tip for Juki Machines: If the machine is noisy, check the hook timing. Juki machines may need timing adjustments more frequently with heavy use.

8. Elna Sewing Machines

Elna machines are known for their compact design and reliability.

Key Maintenance Tips:
• Clean the feed dog area thoroughly, as it tends to collect lint
• Check the needle plate screws regularly, as they can loosen
• Use only Elna-approved bobbins to prevent timing issues

Troubleshooting Tip for Elna Machines: If experiencing thread breakage, check the thread path for any rough spots, especially around the take-up lever.

9. Babylock Sewing Machines

Babylock is known for user-friendly sergers and embroidery machines.

Key Maintenance Tips:
• Clean the loopers thoroughly after each use (for sergers)
• Oil the designated points regularly, following the manual
• Keep the knife mechanism clean and sharp on serger models

Troubleshooting Tip for Babylock Machines: If your Babylock serger is skipping stitches, check the position of the upper looper. It may need slight adjustment.

General Tips Applicable to All Brands:

1. Regular Cleaning:
 • Remove lint and thread debris after every few hours of sewing
 • Pay special attention to the bobbin area and feed dogs

2. Proper Oiling:
 • Follow your machine's manual for oiling points and frequency
 • Use only sewing machine oil, never household oils

3. Needle Care:
 • Change needles regularly, typically every 6-8 hours of sewing
 • Use the correct needle type and size for your fabric

4. Thread Quality:
 • Use high-quality thread to reduce lint and prevent breakage
 • Match thread weight to your project and machine capabilities

5. Professional Servicing:
 • Consider annual professional servicing, especially for computerized models
 • Keep records of servicing and any parts replaced

Troubleshooting Tip for All Brands: If you encounter persistent issues, always consult your machine's manual first. Many common problems have simple solutions outlined in the troubleshooting section.

Remember, while these tips are brand-specific, always refer to your machine's manual for the most accurate maintenance instructions. Each model within a brand may have unique requirements.

Developing a good understanding of your specific machine's needs will help you keep it in optimal condition, ensuring years of trouble-free sewing. Regular maintenance not only prevents issues but also helps you become more familiar with your machine, making troubleshooting easier when problems do arise.

By following these brand-specific tips along with general best practices, you'll be well-equipped to maintain your sewing machine, regardless of its make or model. Happy sewing and maintaining!

Conversion Charts and Tables

Understanding and using the right conversions is crucial in sewing machine repair, especially when dealing with machines from different eras or countries. These charts and tables will help you navigate various measurements and equivalents commonly encountered in sewing machine maintenance.

1. Needle Size Conversion Chart

Different systems are used to measure needle sizes. Here's a conversion chart:

European Size	US Size	Singer Size
60	8	2020
70	10	2032
80	12	2045
90	14	2060
100	16	2080
110	18	2090

Troubleshooting Tip: If you're experiencing skipped stitches or fabric damage, double-check that you're using the correct needle size for your fabric weight.

2. Thread Weight Conversion

Thread weight can be confusing due to different numbering systems:

Weight System	Light	Medium	Heavy
Tex	18-30	40-60	70-90
Denier	160-270	360-540	630-810
Weight (wt)	50-60	30-40	12-20

Note: In the 'Weight' system, lower numbers indicate heavier thread.

3. Oil Viscosity Conversion

Different regions may use different viscosity measurements for sewing machine oil:

SAE	ISO VG
SAE 10	ISO 32
SAE 20	ISO 68
SAE 30	ISO 100

Troubleshooting Tip: Using oil with the wrong viscosity can lead to inadequate lubrication or machine clogging. Always use the recommended viscosity for your specific machine.

4. Metric to Imperial Conversion for Common Measurements

For those working with machines or parts from different systems:

Metric	Imperial
1 mm	0.039 inches
5 mm	0.197 inches
10 mm	0.394 inches

25.4 mm | 1 inch
100 mm | 3.937 inches

5. Tension Disc Pressure Conversion

Some machines use different scales for tension settings:

Numeric Scale	Descriptive Scale
0-1	Very Light
2-3	Light
4-5	Medium
6-7	Medium-Heavy
8-9	Heavy

Troubleshooting Tip: When adjusting tension, make small, incremental changes and test on scrap fabric after each adjustment.

6. Motor Speed Conversion

For understanding motor specifications:

RPM (Revolutions Per Minute)	Stitches Per Minute (approx.)
3000 RPM	750 SPM
5000 RPM	1250 SPM
7000 RPM	1750 SPM
9000 RPM	2250 SPM

7. Timing Angle Conversion

For machines that specify timing in degrees:

Degrees	Needle Bar Position
0°	Needle at lowest point
90°	Needle rising (1/4 turn)
180°	Needle at highest point
270°	Needle descending (3/4 turn)

Troubleshooting Tip: When adjusting timing, always refer to your specific machine's manual, as ideal timing can vary between models.

8. Fabric Weight Conversion

Understanding fabric weight is crucial for selecting the right needle and thread:

Term	Grams per Square Meter (GSM)
Sheer	30-80 GSM
Lightweight	80-150 GSM
Medium	150-250 GSM
Heavy	250-350 GSM
Very Heavy	350+ GSM

9. Power Conversion for International Machines

When using machines from different regions:

Region	Voltage	Frequency
North America	120V	60 Hz
Europe	230V	50 Hz
Japan	100V	50/60 Hz
Australia	240V	50 Hz

Troubleshooting Tip: Always use a proper voltage converter when necessary. Using the wrong voltage can severely damage your machine.

10. Presser Foot Pressure Conversion

For machines with adjustable presser foot pressure:

Numeric Scale	Fabric Type
1-2	Very Light (Chiffon, Lace)
3-4	Light (Silk, Light Cotton)
5-6	Medium (Cotton, Linen)
7-8	Heavy (Denim, Canvas)
9-10	Very Heavy (Upholstery, Leather)

11. Bobbin Size Conversion

Common bobbin sizes and their typical machine types:

Class	Diameter	Height	Common in
15	20.3 mm	11.7 mm	Many domestic machines
66	23.8 mm	11.7 mm	Older Singer machines
M	20.4 mm	8.9 mm	Some European machines
L	23.8 mm	10.4 mm	Large capacity bobbins

Troubleshooting Tip: Using the wrong bobbin size can cause tension issues and damage to your machine. Always use the recommended bobbin type.

These conversion charts and tables should serve as a handy reference when working on various sewing machines or with different materials. Remember, while these conversions are generally accurate, always prioritize the specific recommendations in your machine's manual.

Keeping these conversions readily available can save you time and prevent errors when working on different machines or interpreting instructions from various sources. Consider printing this section and keeping it in your sewing machine repair toolkit for quick reference.

By mastering these conversions, you'll be better equipped to work on a wide range of machines and handle diverse sewing projects with confidence. Happy sewing and repairing!

www.ingramcontent.com/pod-product-compliance
Lightning Source LLC
Chambersburg PA
CBHW071915210526